WORLD BANK TECHNICAL PAPER NO. 401

The Impact of Drought on Sub-Saharan African Economies

A Preliminary Examination

Charlotte Benson
Edward Clay

The World Bank
Washington, D.C.

Technical Papers are published to communicate the results of the Bank's work to the development community with the least possible delay. The typescript of this paper therefore has not been prepared in accordance with the proce-dures appropriate to formal printed texts, and the World Bank accepts no responsibility for errors. Some sources cited in this paper may be informal documents that are not readily available.

The findings, interpretations, and conclusions expressed in this paper are entirely those of the author(s) and should not be attributed in any manner to the World Bank, to its affiliated organizations, or to members of its Board of Executive Directors or the countries they represent. The World Bank does not guarantee the accuracy of the data in-cluded in this publication and accepts no responsibility for any consequence of their use. The boundaries, colors, de-nominations, and other information shown on any map in this volume do not imply on the part of the World Bank Group any judgment on the legal status of any territory or the endorsement or acceptance of such boundaries.

The material in this publication is copyrighted. Requests for permission to reproduce portions of it should be sent to the Office of the Publisher at the address shown in the copyright notice above. The World Bank encourages dissem-ination of its work and will normally give permission promptly and, when the reproduction is for noncommercial purposes, without asking a fee. Permission to copy portions for classroom use is granted through the Copyright Clearance Center, Inc., Suite 910, 222 Rosewood Drive, Danvers, Massachusetts 01923, U.S.A.

ISSN: 0253-7494

Charlotte Benson is senior research fellow at the Overseas Development Institute. Edward Clay is a research fellow at the Overseas Development Institute .

Library of Congress Cataloging-in-Publication Data

Benson, Charlotte.
    The impact of drought on Sub-Saharan African economies : a preliminary
examination / Charlotte Benson, Edward J. Clay.
        p.   cm. — (World Bank technical paper ; no. 401)
    Includes bibliographical references.
    ISBN 0-8213-4180-4
    1. Disasters—Economic aspects—Africa, Sub-Saharan.   2. Droughts—
Economic aspects—Africa, Sub-Saharan.   3. Africa, Sub-Saharan—
Economic conditions—1960–   I. Clay, Edward J.   II. Title.
III. Series.
HC800.Z9D4525   1998
330.967'0329—dc21   97-52167
                                                              CIP

CONTENTS

BOXES

FIGURES

TABLES

ABBREVIATIONS

CCFF	Compensatory and Contingency Financing Facility
CFA	Communauté financière africaine
CILSS	Comité permanent inter-etats de lutte contre la sécheresse dans le Sahel
DFID	U.K. Department for International Development
EWS	Early warning system
GDP	Gross domestic product
GIS	Geographic information system
IGADD	Inter-Governmental Authority on Drought and Development
NGO	Nongovernmental organization
ODA	Overseas Development Administration
ODI	Overseas Development Institute
RMSM	Revised Minimum Standard Model
SADC	Southern African Development Community
SAP	Structural adjustment program
SSA	Sub-Saharan Africa

FOREWORD

This paper arose out of an interest in the broader implications of droughts in African countries and in particular, the droughts in Southern Africa in the mid-1990s.

The current prospects of an El Niño effect in the southern Pacific Ocean, which is forecast to result again in lower than average rainfall in countries in Southern Africa this and next year, makes this paper relevant. The paper provides a number of insights into the impact of drought in Southern Africa in the past and suggests policy options for dealing with future droughts.

It is anticipated that the conclusions from this paper will also be useful for other regions in Africa and the world.

Alan Gelb

Regional Technical Manager

Economic Management and Social Policy

Africa Region

ABSTRACT

Although the physical aspects and agricultural impacts of drought and government and donor responses as well as household coping and survival strategies in the event of drought have been well studied, little research has occurred on either its nonagricultural or economywide macroeconomic impacts in Sub-Saharan Africa. This paper reports the findings of an exploratory study intended as a contribution to filling this gap. A new framework is developed within which to understand the wider economic impacts of drought and explain why some economies are more susceptible to drought than others. Findings include the following:

- Drought shocks have large but highly differentiated economywide impacts. The likely frequency, scale, and character of these impacts depend on the interaction between economic structure and resource endowments, as well as on more immediate short-term effects.

- Counterintuitively, relatively more developed economies in Africa may be more vulnerable to drought shocks than least developed or arid countries, in terms of macroeconomic aggregates and rates of economic recovery. Evidence suggests an "inverted U"–shaped relationship between the level of complexity of an economy and its vulnerability to drought.

- Different regions of Africa are experiencing different long-term climatic trends, implying that different regional strategies are required for mitigation and relief of droughts.

- Although the existence of structural adjustment programs (SAPs) can exacerbate adverse economic and social impacts of drought shocks, such programs also provide a framework within which the broader economywide impacts may be more effectively contained.

- There has typically been little sustained interest in drought mitigation measures on the part of either governments or donors, except in terms of improving food security. There is considerable scope for wider adoption of drought mitigation measures as well as for the incorporation of the risk of drought into economic policies and planning.

- In responding to droughts, financial aid for balance-of-payments and budgetary support should have the highest priority in more complex Sub-Saharan African economies. Large-scale, targeted interventions should be the primary modality of response in simple and conflict-affected economies.

- The complexities of economic structure and resource endowment justify closer exploration of the dynamics of highly drought-vulnerable economies by means of economic modeling.

ACKNOWLEDGMENTS

This paper summarizes the findings of a study by the Overseas Development Institute (ODI) on "The Impact of Drought on Sub-Saharan African Economies and Options for the Mitigation of Such Impacts by National Governments and the International Community," with support from the U.K. Department for International Development (DFID) (formerly the Overseas Development Administration [ODA]) and the World Bank. Provisional findings from the study have been presented at a number of conferences and workshops (Benson 1997, Benson and Clay 1994a, Benson and Clay 1994b, Clay 1994, Clay 1997, SADC 1993b) and at work-in-progress seminars at ODI and in Washington, D.C. in June 1994 and January 1996. The paper has benefited substantially from comments by participants at these meetings and in particular from Harry Walters, Jack van Holst Pellekaan, David Bigman, Xiao Ye, and Essama Nssah. The authors are also grateful to other officials at the World Bank and at DFID, colleagues at ODI, Stefan Dercon (University of Leuven), and Henry Gordon, as well as to colleagues at ODI for their valuable suggestions and comments. Pamela S. Cubberly was responsible for editing and layout of the paper.

CHAPTER 1:

INTRODUCTION

The objective of this paper is to present the findings of a preliminary investigation on a large and complex subject—the impact of drought shocks on Sub-Saharan African (SSA) economies. The intention is to stimulate debate and encourage further systematic analysis at a country and regional level.

The study had its origins in an exploration of the implications of drought for food security policy in SSA (World Bank 1991c), which provoked a fairly consistent response from economists along the following lines. Droughts have serious social impacts, at worst causing famine and associated social disintegration. This poses problems for effective relief, which involves early warning and related preparedness measures to minimize the human costs. It has yet to be established, however, that these events have economic impacts on a scale that justifies economic responses or modifications in policy. This prompted an ODI study to explore the scale and character of the wider macroeconomic impacts of drought in more detail.

The initial working hypothesis for the study was that the economic impacts of drought shocks were most likely to reflect the intensity and duration of drought as a meteorological event. In the absence of any previous comparative study, it was assumed, perhaps reflecting a conventional wisdom, that poorer, low-income countries in SSA, such as Burkina Faso or Ethiopia, were likely to be most severely affected (World Bank 1991c). An initial examination, however, of GDP aggregates for countries widely recognized as having suffered intense drought (as summarized below in chapter 3) and a review of the ways international and bilateral agencies responded to the 1991–92 southern African drought (DHA/SADC 1992, Nowlan and Jackson 1992, SADC 1993b) suggested a more complex picture, involving a counterintuitive relationship between drought and economic structure. It indicated that an economy in the early stages of development could actually become more, rather than less, vulnerable to drought and, therefore, more severely impacted by a shock. In responding to that initial finding, the main thrust of the study has been to explore evidence on the relationship between drought and economic structure systematically. This has involved the elaboration of a typology of countries differentiated by economic structure; an exploration of statistical evidence of the relationships between drought and macroeconomic aggregates, focusing on six case study countries; and a more systematic review of international responses to major drought events, particularly the most recent regional crisis in southern Africa in 1992. In undertaking the investigation, it has been necessary to begin by conceptualizing and defining drought as an economic rather than physical phenomenon (see chapter 2).

Background: Economics of Drought Is Underresearched

Drought is Africa's principal type of natural disaster. Droughts, however defined, are frequent and severe in many African countries as a result of the extreme variability of rainfall in the extensive arid and semiarid areas of the continent and the poor capacity of most African soils to retain moisture. Widely quoted estimates suggest that at least 60 percent of SSA is vulnerable to drought and perhaps 30 percent is highly vulnerable (IFAD 1994). Moreover, parts of the Sahelian belt have been coping with an increasingly dry regime; rainfall is significantly below the norms of the period prior to the 1960s (Hulme 1992).

There has been little research to date on the economywide impacts of drought in Sub-Saharan Africa, either before or after, despite the wider interest of economists in exogenous shocks as well as the potentially serious nonagricultural and macroeconomic impacts of drought.[1] The dearth of investigative studies appears to reflect the fact that drought has typically been perceived as a problem principally of agriculture and, in particular, food supply, with economies automatically and immediately restored to their long-term growth paths with the return of improved rains. Thus, the physical phenomenon of drought, rural household coping mechanisms or survival strategies,[2] and government and donor responses have been extensively studied. Indeed, this body of research has been important in contributing to a broader understanding of drought and to the design of more appropriate relief and rehabilitation responses. However, although this research has also had important implications at an aggregate level for the macroeconomic and financial impacts of drought, it has not directly addressed these two issues.[3]

Similarly, the economywide impacts of drought have often been largely ignored in designing and implementing drought management strategies. SSA governments and the international community have typically responded to droughts by mounting large-scale relief operations, absorbing substantial resources both of the affected countries and of the international community.[4] The primary objective of such operations has invariably been to minimize suffering and loss of human life. Food aid, much of it for use in direct, free distribution programs within the affected areas, has bulked large in relief efforts. For example, some US$4 billion, including associated logistical costs, in food aid and commercial grain imports organized by the governments was provided in response to the 1991–92 drought in southern Africa.[5] In contrast, the importance of nonfood items, such as water equipment, essential drugs, livestock feed, and agricultural inputs, has apparently been less fully recognized (Thompson 1993, DHA/SADC 1992). Moreover, efforts designed expressly to mitigate the impacts of droughts prior to their onset have generally been accorded even lower priority. Such responses partly reflect a widespread failure to perceive droughts as a serious and potentially long-term economic problem.

More positively, the response to the severe 1991–92 drought in southern Africa was slightly different, perhaps suggesting a recent increase in awareness of the economic impacts of drought. Some members of the international donor community displayed a far greater willingness to provide balance of payments support in response to this crisis than to any previous drought, partly reflecting concern about the threat to ongoing economic reform and structural adjustment programs (SAPs), to which donor agencies had committed significant resources.[6] Indeed, the SAPs implied that several of the

---

[1] The comparative lack of research in this area was one of the main findings of a preliminary study at the World Bank (1991b).

[2] For example, Chen (1991), Downing, Gitu, and Kamau (1987), Drèze and Sen (1989), Glantz (1987), Scoones and others (1996), and Sheets and Morris (1974).

[3] For example, Downing, Gitu, and Kamau (1987)—in perhaps the most comprehensive, multidisciplinary review of the impact of a particular drought in Sub-Saharan Africa, focusing on the 1984–85 Kenyan drought—lacks any detailed exploration of the wider or macroeconomic consequences of the drought.

[4] For instance, during fiscal 1984–85 and 1985–86, 7.5 percent of the U.K. aid program was expended on the response to the drought and famine crisis in Africa (Borton and others 1988).

[5] This approximate estimate of the food-related costs of the drought response first appeared in a U.N. *Africa Recovery Briefing Paper* (Collins 1993), subsequently restated in WFP (1993: 22). The costs of drought-related maize and other cereal imports by governments and as food aid was approximately US$2.5 billion in 1992–93 (Mugwara 1994).

[6] For example, in 1992 the World Bank approved a US$150 million Emergency Recovery Loan to Zimbabwe and made additional drought-related modifications to credits of US$50 million for Malawi and US$100 million for Zambia. The U.K. ODA provided both Zambia and Zimbabwe £10 million (approximately US$18

affected economies were already being closely monitored, throwing the economic impact of the drought into sharper focus than had perhaps been the case in earlier droughts in the Sahel and the Horn of Africa. They also underscored the need for a higher level of contingency planning. Nevertheless, many other donors continued to overlook or give only limited attention to the economywide impacts of the drought, such as on foreign exchange availability and government expenditures.

Scope and Purpose

This paper examines the economic impacts of drought in more detail and, in particular, undertakes the following:

- Assess the factors determining the scale and nature of the vulnerability to drought of various types of African economies through an examination of varying drought experiences.[7]

- Provide a preliminary assessment of the impact of drought on both productive sectors and policy targets and objectives.

- Identify appropriate broad drought management strategies that reflect differences in the impact of drought in various types of economies.

This paper also draws on the considerable body of literature on economic shocks more generally. This literature focuses particularly on external shocks, which occur through international goods and capital markets.[8] It provides a wealth of case study materials and insights on factors determining the impact of shocks, including the role of government and private sector responses. It is important to note, however, that the impact of external shocks differs substantially from those of drought, which is an internal shock. Obviously, the immediate direct impact of the former is typically much narrower both in terms of the productive sectors and subsectors affected and of their spatial or geographical impact. In contrast, droughts can affect virtually all aspects of agricultural and other water-intensive activity and impact on a large proportion of households, private enterprises, and public agencies, with far-reaching consequences throughout the economy, as explored in further detail below. Moreover, droughts entail loss of assets in the form of crops, livestock, and productive capital damaged as a direct consequence of water shortages or related power cuts. A further factor limiting the value of the general literature on exogenous shocks is that it focuses particularly on commodity prices, which, as Collier and Gunning (1996) note, commonly involve spikes rather than crashes and so raise a

---

million) in balance of payments support for drought-related imports between March and June 1992. Some donors, for example, Germany, modified existing financial assistance to allow these funds to be used for procurement of drought-related food and other imports. The United States also organized large "blended" packages of support for food imports including export credits, food aid credits and grants to Zambia and Zimbabwe to address the direct balance of payments aspect of the drought (Callihan and others 1994). Other food aid donors also provided a combination of program aid to relieve balance of payments pressures as well as conventional relief for distribution to affected populations.

[7] The concept of vulnerability is subject to many different constructions. It is used here to characterize the scale and duration of potential reductions in aggregate economic activity, sectoral performance, and other economic variables as a consequence of a drought shock.

[8] Krugman (1988) further disaggregates goods market shocks as export, import, and exchange rate shocks, all of which partly depend on price elasticities of supply and demand. Capital market shocks can be disaggregated as shifts in the costs of borrowing, in terms of interest rates, inflation (which in turn influences the volume of extra exports or forgone imports necessary to generate the foreign currency required to service the debt), and the real interest rate. Constraints on borrowing can represent a further shock.

slightly different set of issues and policy decisions. Nevertheless, the literature remains of relevance to this exploration of the economic impacts of drought.

Method of Study

There are considerable methodological difficulties in constructing nondrought counterfactual performance and behavior—that is, in isolating the impact of a particular drought shock from underlying trends and other internal and external factors influencing economic performance. The specification of counterfactual performance and behavior is further complicated by the fact that a drought affects virtually every aspect of economic life. This paper, therefore, adopts an eclectic methodology, using a mixture of quantitative and qualitative analysis. The quantitative analysis is partial, primarily involving an examination of movements around trends, "before and after" impacts, and forecasts compared with actual performance of key economic indicators. Findings from other relevant studies are also drawn on, including some that have entailed detailed econometric modeling.

The ODI study involved a cluster of empirical investigations whose findings and conclusions are reflected in this paper. First, a detailed case study was undertaken of the economic impacts of drought in Zimbabwe since independence in 1980 (Benson 1997, Robinson 1993) complemented by five much briefer desk-based reviews of the impacts of drought in Burkina Faso, Ethiopia, Kenya, Senegal, and Zambia (Benson 1994). Zimbabwe was selected for detailed study for four reasons:

- The country has experienced three major droughts since 1980.

- Its economy is relatively complex compared with most other SSA countries in terms of intersectoral linkages and significant financial flows. This implies that the indirect impacts of drought are potentially important and warrant study because they are less easy to identify. There may also be lessons from the Zimbabwe experience for other SSA economies at earlier stages of development.

- The 1992 drought relief operation in Zimbabwe—and in southern Africa more generally—was far broader than previous responses to droughts in SSA and involved the provision of economic assistance as well as emergency food and other humanitarian relief supplies (Clay 1997).

- Zimbabwe has a relatively good data set, facilitating statistical analysis of macroeconomic and sectoral aggregates.

The five countries selected for desk-based review were deliberately chosen to include examples from the Sahel, East Africa, and southern Africa and also to provide a range of agroecological conditions and economic structures to ensure that the findings of the study were broadly applicable to all drought-prone SSA countries.

Second, because the study was concerned with the relationship between a physical phenomenon—variability in rainfall (precipitation)—and economic activity through an often complex set of possible linkages, it involved a "state of the art" review of the regional incidence of drought, including possible differences in the pattern of rainfall variability and drought risk and whether these phenomena are changing over time (Hulme 1995).

Third, case studies of two specific policy areas were also undertaken: an examination of the interlinking issues of drought and cereal market reform in Kenya (Thomson 1995) and of drought and public expenditure in Namibia (Thomson 1994).

Fourth, in considering policy options for crisis management, mitigation, and preparedness, the considerable evaluative evidence of the southern African drought in 1991–92 was reviewed. That crisis has generated the most extensive body of empirical evidence and policy analysis of any recent major natural disaster. International and bilateral agencies and nongovernmental organizations (NGOs), as well as the governments of affected countries and the regional cooperation agency, the Southern African Development Community (SADC), variously commissioned external evaluations, studies, and meetings to draw out lessons for future policy. It should also be noted, however, that the documentation that is publicly available provides only an incomplete picture. International financial institutions do not publish their assessments of operations involving member governments, although their findings are reflected to some extent in documentation prepared for meetings, (for example, SADC 1993b and World Bank 1995c). Moreover, few NGOs release retrospective assessments of their emergency actions, except where required to evaluate their activities as part of their accountability for grants from public funds (Mason and Leblanc 1993).[9]

---

[9] The selection of studies in the public domain cited in this paper focus particularly on Namibia, Zambia, and Zimbabwe and include: Callihan and others (1994), Clay and others (1995), Collins (1993), Devereux and others (1995), Friis-Hansen and Rohrbach (1993), Hicks (1993), Legal and others (1996), Mason and LeBlanc (1993), Mugwara (1994), SADC (1993a), SADC (1993b), Scoones and others (1996), Seshamani (1993), Thomson (1994), Thompson (1993), Tobaiwa (1993), WFP (1994), and World Bank (1995a).

CHAPTER 2:

POTENTIAL IMPACTS OF DROUGHT ON SUB-SAHARAN AFRICAN ECONOMIES: DIFFERENCES AND SIMILARITIES

A preliminary review of the impacts of drought in Africa suggests that the interaction between drought shocks and the economy are complex, rather than direct and straightforward. But, before exploring that complexity, a working definition of drought that focuses on economic rather than physical or social impacts is required. These conceptual issues are considered in the next section. The main factors determining the nature of the interaction between drought and the economy also need to be taken into account. Those concerning the role of the physical environment, the financial system, and public policy are discussed below before exploring the relationship between economic structure and drought in more detail in chapter 3.

An Economic Definition of Drought

The notion of drought employed for the purposes of this paper is based on a probabilistic conceptualization, encapsulating concepts of meteorological, hydrological, agricultural, and social drought to postulate an economic definition. *Meteorological drought* can be defined as "a reduction in rainfall supply compared with a specified average condition over some specified period" (Hulme 1995). In an African context, this is typically a period of a year or more. *Hydrological drought* pertains to the impact of a reduction in precipitation on natural and artificial surface and subsurface water storage systems, thus, possibly lagging behind periods of agricultural or meteorological drought (Wilhite 1993). *Agricultural drought* is defined as a reduction in moisture availability below the optimum level required by a crop during different stages of its growth cycle, resulting in impaired growth and reduced yields.[10] Finally, *social drought* relates to the impact of drought on human activities, including indirect as well as direct impacts.

There is a circular element in all these concepts of drought, apart from the underlying meteorological phenomenon as a forcing mechanism. Therefore, in undertaking the exploratory investigations reported in this paper, a highly empirical, inductive approach was adopted to establish whether drought could be regarded as a phenomenon of economic significance and, if so, the likely probabilistic nature of the impacts and the implied underlying relationships. The resultant definition of an *economic drought* reflects these investigations. As with agricultural and social definitions, an economic drought concerns the impacts of precipitation-related reductions in water availability on productive activities, including the provision of water for human consumption, rather than water availability in itself. Recurrent, predictable, seasonally low levels or low mean rainfall in arid areas do not constitute drought. Such events are associated with well-established, predictable climatological patterns that occur with a high degree of probability, for example, 80 percent. Thus, these are

---

[10] A number of other factors can also play a role in lowering crop yields, such as reduced input of fertilizer, lack of weeding, the presence of pest and crop diseases, lack of labor at critical periods in the growth cycle, unattractive producer prices, and overall market conditions. It may, therefore, be difficult to isolate the impact of reductions in moisture from that of other factors.

phenomena to which local economies have adapted by selecting less water-intensive types of agricultural and nonagricultural activities and by investing in water storage to smooth seasonal variations in supply.

An economic drought, by contrast, involves low rainfall that is outside the normal expected parameters with which an economy is equipped to cope. Such an event typically results in sharp reductions in agricultural output, related productive activity, and employment. In turn, this is likely to lead to lower agricultural export earnings and other losses associated with a decline in rural income, reduced consumption and investment, and destocking. Meteorological drought may also result in hydrological conditions that have a direct adverse impact not only on irrigated agriculture but also on nonagricultural production, including hydroelectric power generation (which is an increasingly important source of energy in a number of African countries) and certain industrial processes, as well as human water supply. Droughts have additional potential multiplier effects on the monetary economy: the rate of inflation, interest rates, credit availability, levels of savings, the government budget deficit, and external debt stocks. Indeed, the combination of these direct impacts, indirect linkages, and multiplier effects implies that the economywide consequences of a drought may be considerable.

The probabilistic nature of drought as an economic phenomenon should be stressed for a number of reasons. First, it does not involve a simple technical relationship that can be characterized with any certainty, second, it is contingent on the interaction of a meteorological event or anomaly with the changing dynamic conditions of an economy. Third, the impact depends also on the expectations of various economic enterprises—ranging from largely subsistence, peasant households to large private and public corporations, about the probability of drought —and the immediate economic conditions, which in turn are determined by a number of other factors. These considerations suggest a working definition of an economic drought as a meteorological anomaly or extreme event of intensity, duration (or both), outside the normal range of events that enterprises and public regulatory bodies have normally taken into account in their economic decisions and that, therefore, results in unanticipated (usually negative), impacts on production and the economy in general.

According to this definition, a drought can, therefore, be viewed as a form of internal supply-side shock—that is, as a severe disturbance caused by events outside a country's control that have nonmarginal impacts on domestic economic variables.

Environmental Diversity and the Rural Economy

Environmental factors play an obvious role in determining interactions between drought and economic performance. As already indicated, from an economic perspective, a country's vulnerability to drought cannot be viewed merely as more or less synonymous with its aridity. Countries with a high proportion of arid lands are likely to experience frequent and severe drought conditions, thus, precluding significant rainfed agricultural production.[11] Communities in such countries, however, are also likely to have well-developed coping mechanisms; they have adapted to the marginal rainfall conditions through appropriate investments in water resource management and agricultural practices

---

[11] Land areas are widely classified in terms of annual mean potential evapotranspiration (PE), while also taking into account elevation and soil type (Le Houérou and others 1993). Semiarid areas, where cereal production is likely to be subject to limitations because of moisture availability, are defined in terms of annual rainfall falling below an average PE level of 65 percent (R<0.65 PE). Arid areas unsuitable for rainfed cereal cultivation under normal circumstances represent an agroecological zone defined by R<0.35PE (Hulme 1995).

over time. In contrast, predominantly semiarid countries with largely rainfed agricultural sectors are likely to experience nationwide droughts only as an extreme event, as both Zambia and Zimbabwe did in 1991–92. Yet other economies, possibly including Ethiopia and Kenya, according to historical record, are highly unlikely to experience droughts that simultaneously affect most sectors or regions of their agricultural economy. Socioeconomic systems in these latter two categories of countries are, therefore, possibly less well equipped to cope with droughts. Such systems, however, also imply that when droughts do occur they may, in fact, have a larger economic impact in regions less prone to drought. This view of adaptation to drought, also introduces the possibility that only extreme, more improbable, droughts result in significant economywide impacts.

Long-Term Climatic Trends

Interseasonal or annual variations in levels of precipitation may fluctuate around particular trends or cyclical patterns in the long term, rather than remaining within certain stable parameters. The extent of economic vulnerability to drought, therefore, partly depends on the degree to which such patterns and trends are both recognized and taken into account by various economic agents and decisionmakers.

The various regions of Africa have displayed differing climatic trends over the period since decolonization, i.e., broadly, since the early 1960s. Figure 1 compares three regions: the Sahel, East Africa, and southeast Africa.[12] The Sahel has experienced a significant decline in average rainfall levels, defined in terms of a comparison among the three decades before and after 1960. In contrast, for more than a century, the southeast African region of southern summer rainfall has experienced a quasi 18- to 20-year cycle of one relatively wet and one relatively drier decade (Tyson and Dyer 1978). Meanwhile, the East African region is not affected by either significant trends or cyclical patterns in rainfall regime.[13]

---

[12] For purposes of statistical analysis of rainfall patterns, the Sahel is defined as the region lying between 9 and 15 degrees latitude and extending eastward to the Somali border; East Africa is defined as Uganda, Kenya, and Tanzania; and southeast Africa is defined as the region lying between −16 and −26 degrees latitude and east of the 23 degrees longitudinal line.

[13] In trying to understand the various weather patterns, an association between regional rainfall anomalies and the El Niño/Southern Oscillation (ENSO) phenomenon has been established, but this association varies between rainfall zones and explains at most 25 percent of the interannual variation in rainfall (Lindesay and Vogel 1990). Local and global sea surface temperatures also appear to play a significant role, although the precise nature of these relationships is only partially understood to date (Mason and Tyson 1992).

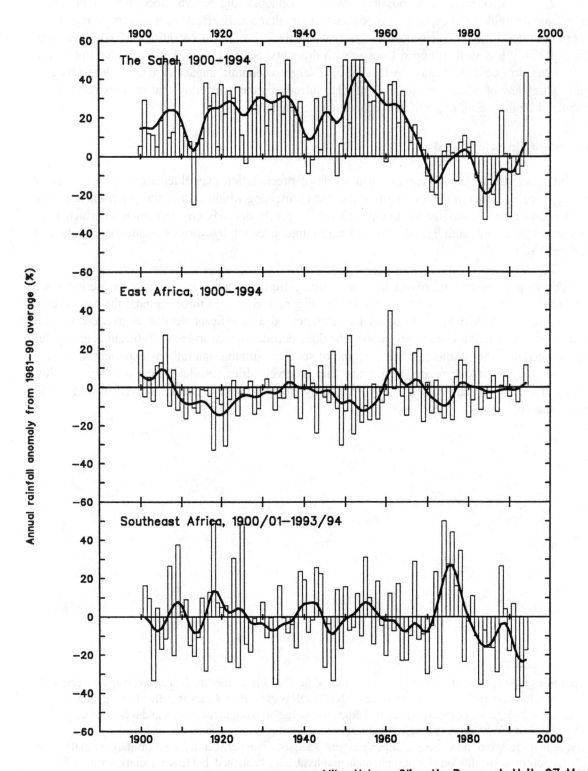

Source: Mike Hulme, Climatic Research Unit, 27 May 1996

The distinct regional climatic differences have potential implications for agriculture, water management more generally, and the possible economic consequences of drought. In some sense the stationary, if random, rainfall patterns of East Africa pose the simplest, most straightforward, drought risk management problem, because the probability distribution of monthly and annual levels of precipitation can be calculated directly from historical records. Farmers, water sector managers, and others have a good grasp of the risks of drought, based on their individual experience (assuming little impact in terms of environmental degradation); thus, from a long-term perspective, they are able to make rational cropping and other decisions based on more or less complete information. In contrast, the Sahelian economies have in effect been confronted with an increasingly unfavorable environment that requires and rewards adaptation through risk-reducing strategies in the rural economic systems.[14] Meanwhile, the quasi-cycles of southeastern Africa may pose particularly difficult problems in making appropriate decisions on hazard management and investment. Decisions based on rainfall or hydrological information, such as river flows or recharge of aquifers for only part of one or more full cycles (10 to 15 or 30 years) or which assume random distribution of drought years rather than taking account of the actual rainfall cycle, could involve greater than intended exposure to drought hazards. The assessment of drought risk in southern Africa is further complicated by as yet highly tentative indications at the level of regional analysis of small future reductions in mean annual rainfall and an associated increase in the expected frequency of extreme events such as the "1 in 50 years" drought of 1991–92 (Hulme 1996).

Role of Financial Systems

The evolution of financial systems is not linked in any simple way with levels of GDP. The size and structure of the private financial sector, involving banks, other intermediaries, and elements of a private capital market, however, will have potentially significant implications for the way an economy responds to drought and other economic shocks and the types of adaptive behavior available to individual households, the public sector, and public authorities. Following a drought shock, demand for agricultural credit and defaults on previous loans are both likely to rise. In countries where extensive formal savings institutions for small savers exist, substantial internal flows of resources may occur, as those in less-affected areas transfer remittances to their relatives in greater need, thus, effectively spreading the household impacts of drought more evenly. Similarly, the CFA franc facilitates transfers between countries within the zone, making remittances an important source of income to drought-affected rural family members (Reardon and Taylor 1996). Affected enterprises are likely to face reductions in their financial balances and, therefore, to seek further loans or extended credit. Increased public sector budgetary pressures and parallel pressures on the foreign exchange account because of increased public sector imports could also place additional strains on the financial system, depending on how such gaps are financed. Increased pressure on financial markets may be partly offset by reduced rates of investment borrowing and by possible declines in private sector demand for imported raw materials and intermediate goods. There is likely, however, to be some remaining financial gap, which, depending on how it is met, will have varying implications for the impact of the drought on the economy.

Role of Public Policy

At independence, most African economies inherited public enterprises with responsibility for agricultural marketing, electricity supply, water supply, waste disposal, and transport. Public institutions were also heavily involved in formal human resource development and social welfare provision through

---

[14] For example, Davies (1995) provides substantial evidence of such adaptation, which, at the level of household and specific groups, may imply transitional stress and impoverishment.

education, health, and other social services. Public sector involvement in both agricultural and nonagricultural sectors increased in the early post-colonial period, bringing many problems of economic efficiency and public finance, which are now widely recognized. Subsequent economic reforms and the growth of both informal and regulated sectors have further modified and sometimes reduced the role of the public sector in the production and distribution of goods and the provision of public services. These gradual changes imply that the role of the public service and thus the effects of a drought shock have varied both over time and among countries. For example, parastatal marketing depends to differing and changing degrees on large-scale commercial production and the surpluses of small-scale, self-provisioning peasant farmers. The effects of drought shocks may be amplified where marketing depends on the production surpluses of small-scale farmers, as discussed in chapter 4.

As the wider literature on exogenous shocks indicates, the nature of prevailing policies and particular policy responses can also play a critical role in determining the precise nature and scale of impact of an exogenous shock. Commodity price shocks, for example, can generate instability in both government revenue and the income of private agents (Collier and Gunning 1996). Two decades ago, it was widely thought that the government should, therefore, play a custodial role, seeking to smooth private incomes in the event of positive price shocks by transferring the volatility to itself.[15] Subsequent research has indicated that various factors have played a role in determining the quality of management of windfall incomes, including prevailing policies.[16] More generally, governments have typically made a range of errors in mismanaging windfall incomes, in part because of some tendency for government intervention to cloud information about the precise nature or expected duration of the shock, which in turn has confused the private sector about an appropriate response.

Similarly, the precise nature of the impact of droughts depends partly on existing policies and drought-related policy responses, be they to ensure that a government's budgetary targets are reached despite additional drought-related expenditure or that prior export commitments are met. These issues are explored in further detail in chapter 4.

There is an important advantage in addressing the effects of a drought compared with many other forms of shock—namely, that its duration and temporary nature are transparent. This effectively provides public and private agents with relatively good information about appropriate behavior. There is also, however, the danger that, as an event that is widely perceived to be beyond a government's control, a drought will become the automatic scapegoat for any contemporaneous economic woes and will thereby delay the implementation of appropriate measures to address other, possibly more fundamental economic problems. For example, by attributing the consequences of underlying economic difficulties to the impact of drought, the Zimbabwe government may on occasion have adopted inappropriate policies that adversely affected long-term economic performance. The 1982–84 drought coincided with a period of severe balance of payments difficulties. Davies (1992) argues that the government initially viewed these difficulties as

---

[15] This belief was based on one or more of four propositions: *(a)* that the government is more farsighted than the private sector and so more likely to conserve a windfall, *(b)* that the social cost of taxing windfalls is particularly low because the windfall was not expected, *(c)* that if windfall income accrues to the private sector it can cause inflation, and *(d)* that if windfall income is not placed in government reserves, it will generate an appreciation of the real exchange rate (Collier and Gunning 1996). Subsequent research has proved that all four propositions are poorly founded.

[16] For example, the Ghanaian government increased its foreign exchange reserves during the cocoa boom by almost the entire amount of the windfall, possibly as an inadvertent consequence of the country's elaborate system of foreign exchange rationing (Collier and Gunning 1996).

temporary, arising as a result of the drought. Rather than seeking to address the underlying problems, it, therefore, employed short-term quick-fix solutions in the form of import controls. The 1991–92 drought could be similarly blamed for the difficulties arising during Zimbabwe's recently adopted adjustment program, thereby helping the government to sustain the reform process.

To sum up, a preliminary review suggests that the economic implications of a drought shock depend on a complex set of environmental and economy-specific factors. Nevertheless, the available evidence suggests that certain features of economic structure are overwhelmingly important in determining drought vulnerability and that these can be captured in a relatively simple typology of country situations, which is presented in the following chapter.

CHAPTER 3:

ECONOMIC STRUCTURE AND THE IMPACT OF DROUGHT

As a preliminary exercise in exploring the impact of droughts in more detail, the GDP and agricultural GDP performance of a number of countries during years of widely recognized severe drought was examined. Several SSA economies, including South Africa, as well as two other drought-prone countries, India and Australia, were considered in the analysis. The results of this comparative analysis are indicated in figure 2. This figure plots the change in real GDP attributable to changes in agricultural GDP against the total change in GDP, comparing years of severe drought with performance in the previous year. For example, a 50 percent fall in agricultural GDP in an economy in which agricultural GDP had accounted for 20 percent of total GDP in the predrought year would translate into a 10 percent fall in GDP attributable to the decline in agricultural GDP.

The results generated threw up some apparently strange anomalies, raising a number of questions. Why, for example, was Burkina Faso located near the (x,y) axis, implying only a small drought-related change in GDP, whereas Niger lay toward the far left of the figure indicating compensatory growth elsewhere in the economy (uranium mines) that partly offset the impact of the drought? Why did some countries lie to the right of the 45 degree axis, implying that droughts were largely agricultural phenomena, whereas others were located to the left of the line, suggesting that drought also had adverse impacts on nonagricultural sectors? And why had Zimbabwe's positioning on the scatter diagram changed so substantively between the 1982–83 and 1991–92 droughts?

These preliminary investigations suggested a complex relationship involving differences of economic structure determining the severity of drought impacts. Contrary to initial expectations (World Bank 1991c), these explorations suggested a counterintuitive relationship: an economy in the early stages of development may become more vulnerable to drought and more severely impacted by drought shocks.

Typology of Country Situations

To focus more clearly on the structural features of an economy that mediate the effects of drought as a climatic and hydrological event, it has been found useful to develop a typology distinguishing four distinct country situations in terms of the impact of drought[17]:

---

[17] This typology builds on one initially presented in Nowlan and Jackson (1992). The authors proposed a typology of subsistence, developing, and complex economies as a way of explaining the need for different forms of international response to assist the drought-affected economies of southern Africa in 1992.

Figure 2: Impact of drought shocks on GDP, the role of performance in the agricultural sector

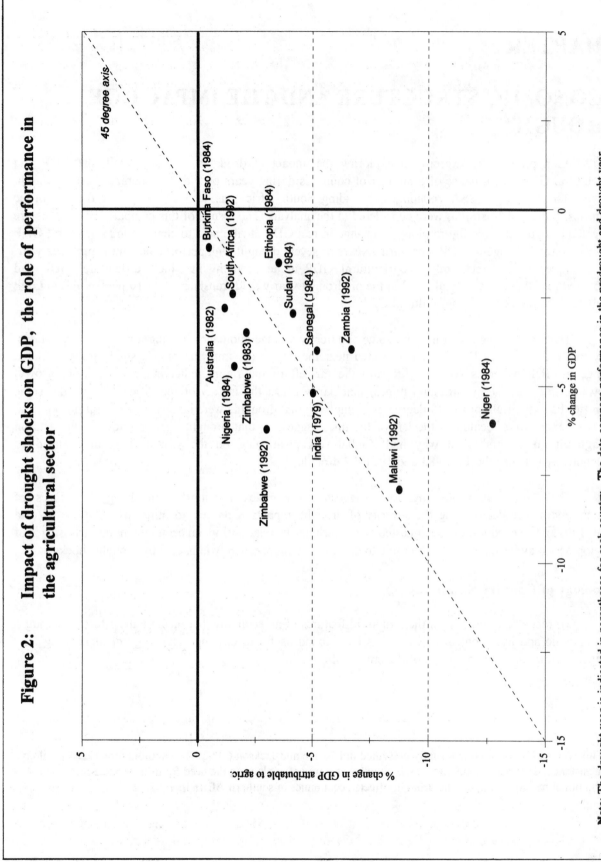

Note: The drought year is indicated in parentheses for each country. The data compare performance in the predrought and drought year.

Source: Based on data derived from World Bank, 1995, "World Data 1995: World Bank Indicators on CD-ROM", Washington D.C.

- *Simple economies.* These are predominantly rainfed agricultural and livestock semisubsistence economies with a limited infrastructure, low levels of per capita income and high levels of self-provisioning in the rural population. To the extent that a modern sector exists, it has few links with the agricultural sector. Commodity and factor markets may be incomplete and poorly integrated, at least at a national level. These characteristics may be further exaggerated where conflict results in a disarticulation or loss of economic complexity in both the real and monetary systems (Clay 1997).

- *Intermediate economies.* These are more diversified economies with economic growth occurring via the development of a labor-intensive, low-technology manufacturing sector, typically dependent on domestically produced, renewable natural resources and imported inputs and capital equipment, but with natural resources still representing a relatively important part of export earnings.

- *Complex economies.* These are developed economies with a relatively small agricultural sector and proportionately small forward and backward linkages between the agricultural sector and other water-intensive activities and the rest of the economy. To date, no SSA country could be viewed unambiguously as a complex economy.

- *Dualistic economies with large extractive, minerals sectors.* These are economies consisting of a "traditional" low (labor) productivity rural economy, entailing a high level of self-provisioning, which coexists alongside a "modern" economy involving an export-oriented sector closely integrated with a service sector but which is relatively immune to performance in other sectors. Drought will affect only part of a dualistic economy unless the export-oriented extractive sector is water intensive. Such dualism is a characteristic feature of many SSA economies, including drought-prone Botswana, Namibia, and Niger, as well as South Africa, which has the highest GDP per capita in the region.

The varying characteristics of these four types of economies are indicated in more detail in table 1 below.

In reality, the typology is highly simplified. There is, in fact, a continuum of cases, as specific structural characteristics are not unique to economies lying within a particular category. For example, Australia is a complex economy with a large mineral sector, whereas India, although falling into the intermediate category, has substantial industrial sectors. Regional dimensions can also be important. For instance, Namibia is not particularly vulnerable to the direct impacts of drought. It is, however, being closely integrated into the wider SADC region where drought shocks can have profound impacts, with ramifications for Namibia.

Another major caveat with regard to the typology is that it ignores the fact that economies are dynamic rather than static entities and that the impact of a specific drought shock is, therefore, partly time dependent. In particular, economic restructuring, financial deepening, urbanization, and increasing regional integration have all been important trends in SSA in recent years. Regional integration has been particularly rapid in southern Africa, following the end of apartheid. Conflict-related factors can also play a role, as already indicated. Just as conflict involves a loss of complexity in Somalia and Sudan, for example, so, after periods of extended conflict, Ethiopia and Mozambique are acquiring characteristics of more complex, intermediate economies, thus, increasing their vulnerability to drought (see below).

Table 1:   Characteristics of the four country typology

	Simple	Intermediate	Complex	Dual (a)
Per capita income	Low income country	Low/lower-middle income country	Higher-middle/higher income country	?
Importance & nature of the agricultural sector	Predominantly rainfed, accounting for 20%+ of GDP & of 70%+ employment	Rainfed/irrigated, accounting for 20%+ of GDP & 50%+ of employment	Predominantly irrigated accounting for under 10% of GDP & under 20% of employment	?
Inter-sectoral linkages to & from water-intensive sectors	Weak	Intensive	Diffused	Weak
Engine of growth	Agricultural sector	Agricultural/non-agricultural sectors	Non-agricultural sectors	?
Level of infrastructure & market development	Weak	Medium	Strong	?
Stage of financial market development	Weak, largely informal	Medium	Strong	?
Vulnerability to primary export commodity price shocks	High	Medium	Low	High
Trade regime	Semi-open	Semi-open/open	Open	?
Quality of management of natural resources	Poor	Fair	Good	?
Spatial impact of drought	Largely rural, confined to immediately affected regions	National, affecting urban as well as rural population	Largely rural, confined to immediately affected regions	Rural
Main source of funding of drought relief programme	Donors	Domestic government/donors/private sector	Domestic government/private sector	?

a Dual economies include low (eg, Niger) and middle income (eg, Botswana and Namibia) economies and so display a range of characteristics with regard to some aspects of their economies

Nevertheless, the typology remains useful in focusing attention on how drought shocks interact with different types of economic structures. The precise nature of the impact of drought in the different country typologies is explored in further detail below.

Impacts of Drought

The initial direct or physical effects of drought on the productive sectors are similar regardless of the type of economy, although the relative and absolute magnitudes of each shock will depend on specific country characteristics. A decline in rainfall has an initial adverse impact on the agricultural and livestock sectors, on hydroelectric power generation, and on other water-intensive activities. Domestic availability of water is also restricted; this has implications for health and household activities, including the time required to collect water. Increased competition for more limited water resources may also occur, possibly necessitating controversial policy decisions (for example, on the relative allocation of water resources to hydroelectric power generation and irrigation).

Drought shocks then have a range of second round and subsequent effects, the nature and magnitude of which depend on particular country circumstances, as explored in further detail below. For example, declines in relatively water-intensive output may constrain the productivity of related sectors and subsectors as well as of irrigated agriculture. Some loss of earnings may occur as labor is laid off, overtime bans are imposed, and shorter shifts are worked, in turn reducing demand in the economy. Shortages of food and other goods may drive up the rate of inflation, although partly offsetting losses in agricultural value added. Budgetary and external trade balances may deteriorate, whereas external debt shocks increase. The broad schema of these impacts is indicated in figure 3.[18]

Drought may also exacerbate income inequality, undermining efforts to achieve equitable and sustainable development (see box 1). Although short-term changes in patterns of income distribution are difficult to demonstrate quantitatively, the increasing priority accorded by the international community to poverty reduction is reflected in more systematic research on the effects of economic shocks, including drought, on poorer groups and regions (Reardon and Taylor 1996).

Simple Economies

The economic impact of drought is largely felt via its direct impact on the agricultural sector. This is reflected in substantial percentage declines in GDP, agricultural exports, and employment opportunities as well as in widespread sale of assets. Because of the relative importance of the agricultural sector, the overall economic impact of drought may be particularly great. Severe drought also results in widespread nutritional stress, higher morbidity, and, possibly, higher mortality. However, because of weak intersectoral linkages, a high degree of self-provisioning, relatively small nonagricultural sectors, and often poor transport infrastructure, the multiplier effects through the rest of the economy are fairly limited; they largely occur through a decline in consumer expenditure. The effects of drought are likely to be concentrated in the rural economy.

---

[18] The flow chart deliberately abstracts from the "social" consequences of drought. Such effects are not directly considered because of the methodological difficulties entailed in trying to incorporate them.

Figure 3: Transmission of a drought shock through an economy

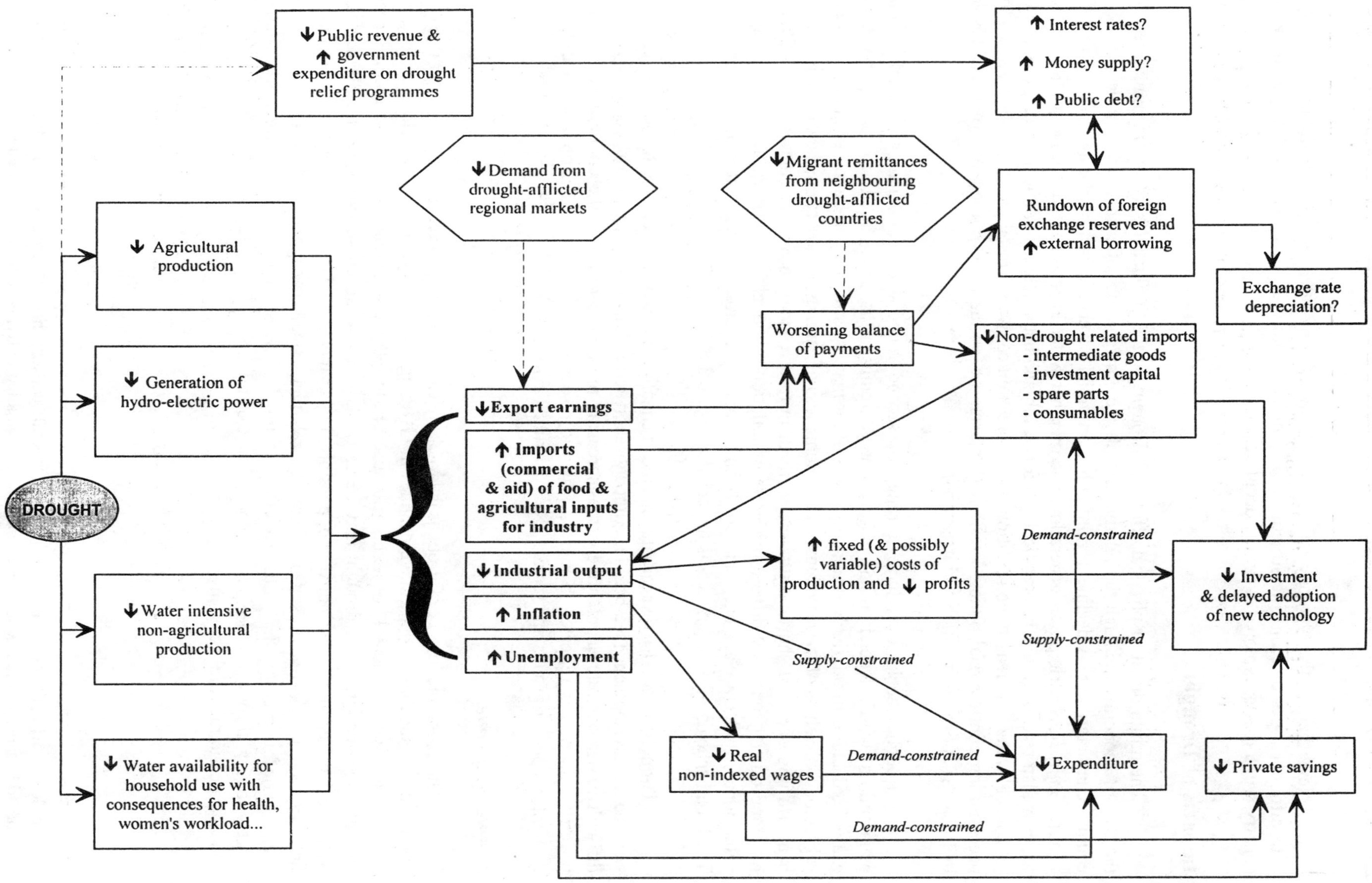

Box 1: Impact of Drought on the Distribution of Income

Drought shocks can have differing impacts for various socioeconomic and geographically located groups, with potentially significant implications for the pattern of income distribution, as in the following examples:

- Differential impact of drought on employment opportunities among sectors and subsectors; certain subsectors, such as mining and some service industries, are relatively immune (assuming an adequate electricity supply) (Robinson 1994).

- Differential impact of drought on the supply of and returns to labor in the formal and informal sectors. For example, Berg (1975) argues that rural-urban migration during the 1972–74 drought in the Sahel resulted at first in increased movement of labor into the urban informal sector. This reduced returns to labor in this sector, whereas wages in the formal sector, which were protected by statutory minimum wage regulations, increased substantially.

- Varying degrees of vulnerability within sectors. Within the agricultural sector, peasant farmers are likely to be more vulnerable to drought because of their predominant involvement in rainfed agriculture and their greater concentration in less favorable climatic regions. As a consequence, they are likely to face much greater relative loss of assets, thus, widening disparities between small- and large-scale commercial producers. For similar reasons, pastoralists with limited numbers of livestock may be affected worse than large-scale commercial cattle farmers (Devereux and others 1995; Scoones and others 1996).

- Differential impact of drought between urban and rural areas. Drought may entail a two-way redistribution of income between rural and urban households, although rural to urban flows probably dominate. These occur most fundamentally via the greater deterioration in terms of trade suffered by rural households and the potentially severe impact of drought on their main source of earnings. However, where some members of a family have migrated to towns and cities, these losses may be partly offset by increased flows of urban-rural remittances (Hicks 1993).

- Differential impact of drought among regions depending on rainfall levels, catchment areas of major supplies of water, and the relative importance of water-intensive activities in each region. Ease of access to relief supplies relative to need may also vary among regions (Downing and others 1989).

- Gender dimensions in the event of death. If a male head of household dies, according to customary laws among, for example, the Tonga of Zambia (Ainsworth and Over 1994), the widow has no entitlement to any of the household's possessions. More generally, female-headed households (in which there is no absent male sending remittances) often fare worse, in part reflecting their typically smaller resource base (Bonitatibus and Cook 1996).

Drought shocks may also have long-term implications for income distribution—for example, owing to the wide-scale sale of assets by low-income households at prices significantly below the nondrought norm. This, in turn, for instance, can have implications for demand for education in some countries. For example, in Zimbabwe goats are often kept as a form of savings to pay for secondary education (Hicks 1993). However, a number of households were forced to sell their goats to sustain short-term levels of consumption in the aftermath of the 1991–92 drought shock.

Recovery is relatively fast. Assuming the timely availability of sufficient seed, draft animals or agricultural machinery, other agricultural inputs and tools, and predominantly annual cultivation cycles, the restoration of good rains can restore levels of GDP to predrought levels almost immediately, although recovery will be slower in the cases of livestock production and of sugar cane, coffee, and other crops with a multiyear production cycle. However, drought may leave a legacy of higher levels of internal and external debt, larger balance of payments deficits, and reduced and less equitably distributed capital assets, such as livestock and household items.

Intermediate Economies

The effects of drought are diffused more widely through the economy, reflecting greater overall integration and stronger intersectoral linkages than displayed in simple economies. Droughts affect the (larger) manufacturing as well as agricultural and livestock sectors, as the lower domestic production of agroprocessing inputs reduces nonagricultural production, while forcing up input costs. Intermediate goods are also likely to form a larger share in total imports, implying that any drought-related import squeeze will have additional, multiplier implications for domestic production. Meanwhile, consumer purchasing power is likely to decline owing to some combination of higher food prices, job losses in both agricultural and nonagricultural industries,[19] nominal wage freezes, and the reduced availability and higher cost of credit. Falling markets, rising input costs and possibly higher interest rates can also result in delayed investment in new capital and technology, with long-term growth implications.

In the aftermath of a drought, as in a simple economy, the agricultural sector will recover relatively quickly. However, recovery of the manufacturing sector may be slower because of the combined impacts of such difficulties as continued input shortages, in turn reflecting ongoing foreign exchange problems and only a slow pickup in demand (see box 2).

The structure of financial sectors and government financial policy are also likely to be more important in shaping the impact of a drought shock than in a simpler economy. The government is itself likely to meet a larger share of the costs of the relief efforts, rather than relying almost entirely on international assistance. This will be financed by some combination of the reallocation of planned expenditure, government borrowing, and monetary expansion, with various indirect long-term implications, as discussed in further detail in chapter 4. Large interannual fluctuations in economic performance, such as those that drought causes, can also create economic management difficulties, for example, in controlling public expenditure.

Intermediate economies typically have more developed, economywide financial systems for the flow of funds, including small-scale private savings and transfers, which also diffuse the impact of drought more widely. For example, the transfer of remittances from urban- to rural-based members of households was facilitated by the well-articulated system for small savings in Zimbabwe in the aftermath of the 1991–92 drought. This mitigated the impact of the drought on the rural areas but at the same time effectively spread its impact more widely, including into urban areas (Hicks 1993).

---

[19] Reductions in demand for labor are unlikely to be fully reflected in official statistics, because these effects are partly addressed through restrictions on overtime, shorter working shifts, and laying off of casual labor. Contract labor may also be laid off but permanent labor forces may be protected by stringent employment regulations such as significant redundancy payments, which prevent large declines in formal sector employment.

Box 2: Impact of Drought on the Manufacturing Sector in Zimbabwe

Largely as a result of the 1991–92 drought, Zimbabwean manufacturing output declined by 9.5 percent in 1992. The drought alone led to a minimum 25 percent reduction in the volume of manufacturing output and a 6 percent reduction in foreign currency receipts from manufactured exports or a 2 percent reduction in total export receipts. The textiles (including cotton ginning), clothing and footwear, nonmetallic mineral products, metal and metal products, and transport equipment subsectors were particularly badly affected. The drought impacted on the manufacturing sector in a variety of ways:

- *Water shortages.* Most municipalities imposed rationing; particularly severe water shortages were experienced in the cities of Mutare, Chegutu, and Bulawayo.

- *Electricity shortages.* Reduced hydroelectric production (see box 4) resulted in load shedding, rationing, and increased electricity tariffs affecting the whole country. Load shedding imposed particular costs on subsectors with batch or continuous processing, whereas the system of rationing discriminated against smaller manufacturers.

- *Input supply difficulties.* Shortages of agricultural inputs to the manufacturing sector were experienced, with adverse implications for most agroprocessors. However, larger food processing companies, such as grain millers, increased production as imports were channeled through urban plants rather than processed in smaller rural plants. Similarly, the meat processing industry faced increased supply of inputs as the drought forced up slaughtering rates.

- *Reductions in demand.* Demand for both agricultural inputs and other basic consumer goods such as clothing and footwear fell, partly due to the contractionary effects of the drought and an ongoing structural adjustment program, as well as increased penetration of the Zimbabwe market by competitive imports following recent trade liberalization.

- *Macroeconomic conditions.* Higher government domestic borrowing, in part to finance drought-related expenditure, higher rates of inflation, and higher nominal interest rates created an unfavorable operating environment. Subsectors in which working capital requirements had increased sharply because of parastatal price rises (e.g., for steel) were particularly severely affected.

Partly as a consequence of the drought, the International Finance Corporation identified the Zimbabwe Stock Market as the worst performer of fifty-four world stock markets in 1992; it had a 62 percent decline in value. Although increased costs of production were partly passed on to consumers, manufacturers faced a deterioration in their financial viability.

Source: Robinson (1993).

In addition, although the extent of absolute poverty is likely to be lower in an intermediate than simple economy and the nature of household vulnerability to drought is likely to have changed, it may not necessarily be much reduced. There will still be some subsistence farming. Furthermore, vulnerability does not depend solely on levels of poverty in nondrought years but also on the ability of households to cope with drought and other adverse conditions. Development involves some degree of specialization, a decline in self-provisioning and fuller integration into markets and financial systems, altering but not necessarily reducing a household's vulnerability to drought (Clay 1997).

Complex Economies

The impacts of a drought are relatively easily absorbed in more complex economies, in part reflecting the typically smaller contribution of the agricultural sector to GDP, exports, and employment (see box 3).[20] Water resources are also likely to be better managed. The likely scale of impact of drought shocks in such economies is illustrated by Australia's experience during the 1982 and 1994–95 droughts. In 1982 agricultural output, which had represented 5.2 percent of GDP in 1981, fell by 29 percent, but GDP declined by only 2.8 percent (Hogan and others 1995). In 1994–95, although farm production alone fell by 9.6 percent in gross value terms, simulations indicate that this latter drought reduced the forecast rate of economic growth from 6.4 to 5.2 percent in 1994–95 and from 4.4 to 4.0 percent in 1995–96. The lagged impact of the drought reflected lower wool and cattle production, in turn, due to a reduction in herd size.

Complex economies are typically both more open and have fewer foreign exchange constraints, facilitating the import of normally domestically sourced items in the event of a drought, without forcing a decline in other imports. Real exchange rates may appreciate marginally as the price of agricultural commodities increases, but this will probably be temporary.

Average per capita incomes are higher and food items account for a smaller percentage share of total household expenditure, implying that, even if the prices of drought-affected food products rise, the purchasing power of most groups will not be significantly altered. Thus, the scale and cost of relief programs will be limited, avoiding any substantial increase in government domestic or external borrowing. However, that small segment of the population that is affected—largely farmers in drought-affected areas—may be severely hurt in terms of loss of income, assets, and savings.[21]

Dualistic Economies with Large Extractive Sectors

Some drought-prone economies in SSA exhibit a high degree of dualism; they have a large capital-intensive extractive sector that features significantly in the trade account but is weakly linked with other sectors of the economy. Some African economies have achieved relatively high per capita levels of GDP through development of such sectors, underlining the fact that there is no linear model of development.

Unless the extractive sector is water intensive and fails through lack of investment or poor management to insulate itself from fluctuations in water supply, the economic impact of drought in such economies is likely to be limited to variability in the agricultural sector and have only limited multiplier effects. Thus, the macroeconomic impact of drought appears small, similar to what is true in a complex economy. But this impression is deceptive, overlooking the profound impacts

---

[20] For example, California experienced six consecutive years of drought between 1987 and 1992. The 1990 drought alone resulted in estimated agricultural losses of US$450 million. However, only one subsector of the urban economy—the urban landscape gardening industry—was seriously affected (California Department of Water Resources 1991).

[21] For example, Purtill and others (1983) estimated that farm incomes for Australia's "broadacre" properties fell by an average of 45 percent during the 1982–83 drought, with declines as high as 96 percent in Victoria. Debt held by drought-afflicted farms increased fourfold during the period between June and November 1982. The drought also resulted in a 2 percent fall in employment nationwide.

Box 3: Modeling the Economic Impact of the 1992 Drought in South Africa

The agricultural sector accounts for a relatively small share of GDP in South Africa. In 1991, for example, it accounted for 6.6 percent of GDP, a level similar to that in the United States and Australia. However, as compared with other complex drought-prone economies, the sector provides a relatively important source of employment. During 1985–88, 13.6 percent of the South African workforce was employed in agriculture compared with 5.5 percent in Australia and 3.0 percent in the United States. Thus, drought might be expected to have a relatively greater impact on, for example, domestic private savings and demand.

The Reserve Bank of South Africa developed a macroeconomic model to examine the impact of the 1992 drought (Pretorius and Smal 1992). The model is based on the Keynesian income expenditure approach. It incorporates the supply side of the economy through a number of equations modeling the value added by different output sectors; supply side constraints are determined by full employment of the labor force and fixed capital stock in a neoclassical production function. The agricultural multiplier was calculated at 1.6 for the country overall, although it could be higher in predominantly agriculture-oriented rural areas.

The predictive powers of the model were broadly correct after adjusting for the fact that the decline in agricultural GDP was twice the level assumed in running the model. The simulation results indicated that a 14 percent decline in agricultural sector value added would result in a 1.8 percent decline in GDP, of which 1 percent would be direct and 0.8 percent indirect impacts. In reality, agricultural GDP fell by 27 percent whereas GDP declined by 2.4 percent. The decline in agricultural GDP directly accounted for a 1.5 percent fall in GDP. The direct impact of reduced rainfall on the manufacturing sector was comparatively limited; output declined by only 3.3 percent because past investment in urban water supplies effectively assured the availability of water. The other computed effects included falls of 1.8 percent in real disposable income, 0.5 percent in consumption expenditure (due to lower disposable income and higher food prices), 5 percent in gross domestic saving, and an incremental rise in the rate of inflation of 0.8 percent. Consumer expenditure actually fell by 0.9 percent, and gross domestic savings fell by 8.4 percent. The model also estimated a 0.5 percent decline in gross domestic investment, but a much higher fall of 4.2 percent was experienced.

The drought necessitated maize imports totaling some R1,725 million (US$604 million) between April and December 1992 alone; imports continued into 1995. Maize export earnings fell by an estimated R365 million, with further declines in other agricultural exports as well as an estimated R335 million drop in exports from related sectors. However, lower domestic demand as a consequence of the drought was expected to reduce nonagricultural imports by R1,200 million. Meanwhile, the drought was estimated to have resulted in the loss of 49,000 agricultural jobs and 20,000 formal sector jobs in nonagricultural sectors. Farmers' indebtedness was also expected to have risen, cutting some farmers off from access to further credit.

within the agricultural sector on which the majority of the population remains dependent. Droughts can result in a considerable intensification of food insecurity, water-related health risks, and loss of livelihoods in the agricultural sector. However, in contrast to simple economies, the broad revenue base and the scope for maintaining financial stability provided by taxing the extractive sector offer considerable opportunity for countervailing measures.[22]

---

[22] For example, in Botswana in 1982–87 (Drèze and Sen 1989) and in Namibia in 1992–93 (Thomson 1994), the macro aggregate and trade account effects of drought were modest and governments had the resources to finance substantial relief programs. The Botswana example is particularly exceptional, because it has managed its diamond revenue extremely well, treating high as well as low prices as temporary shocks.

The "Inverted U" Relationship

In summary, this chapter has proposed that features typical of a simple economy effectively contain the economic effects of drought, with the impact largely felt at the rural household level and within the informal economy. As an economy develops, with diversification into the manufacture of technologically simple products utilizing domestically produced raw materials, growth in financial and commodity markets, and an expansion of the monetized consumption base and nonagricultural sectors, its economic vulnerability to drought shocks initially increases. Agricultural earnings increasingly take the form of cash as households grow a smaller proportion of crops for their own consumption, again spreading the impact of any downturn through the economy. The accompanying specialization of labor and the breakdown of community and extended family ties may reduce the ability of households to adapt to temporary shocks. In the later stages of development, vulnerability to drought shocks then declines again as the agricultural sector becomes decreasingly important both in GDP and as a source of employment, also implying weaker forward and backward linkages between the agricultural sector and the rest of the economy. Such economies are also relatively more open and do not face major foreign exchange constraints, ensuring that any domestic shortfalls can be met through imports without affecting other trade flows.

This conceptualization of drought and economic structure contained in the stylized four-country typology together with the evidence presented in figure 2 suggest a challenging hypothesis—that there is an underlying "inverted U"-shaped relationship between the macroeconomic impact of drought and the overall stage of economic development of a country.[23] In other words, the economic impact of drought increases during the earlier stages of development before declining as an economy becomes more developed rather than, as conventionally assumed, continuously declining as an economy becomes increasingly complex. In an African context, this hypothesis would appear to have become increasingly relevant over time. Economic thinking has shifted away from highly regulated to more market-oriented policies, implying that the impacts of drought as well as other exogenous shocks are increasingly felt in the domestic economy. Moreover, some African economies have achieved relatively good progress in their economic development over the past two or three decades, making the experiences of intermediate economies increasingly relevant to some countries. Higher oil prices have resulted in increased investment in hydroelectric power generation, whereas rising populations have placed ever greater demands on limited water resources.

[23] Kuznets' "inverted U" hypothesis was based on evidence that relative income inequality rises during the earlier stages of development, reaches a peak, and then declines in the later stages (Kuznets 1955).

CHAPTER 4:

DROUGHT AND THE ECONOMY: SOME PROVISIONAL FINDINGS

This chapter presents the findings of an exploratory investigation of the evidence on the factors determining the economic impact of drought shocks (chapter 3). Evidence is examined on the overall impact of disasters, whereas a range of both exogenous, externally determined, and domestic factors lying within the control of policy makers and other economic agents that can influence the outcome of a severe drought shock are also considered. The interactions of drought shocks and structural adjustment are explored separately in chapter 5.

The Role of Economic Structure: The Southern African Drought of 1991–92

The initial evidence on a complex interaction between drought and economic structure and a possible "inverted U," as presented in chapter 3, was merely suggestive. The process of selecting countries for inclusion in figure 2 was subjective and unsystematic, reflecting the investigators' awareness of recent major drought events. A comparison of a set of economies simultaneously impacted by the same event would provide a more rigorous test of the validity of the relationship. The 1991–92 southern African drought comes closest to providing such a test, recognizing that this drought impacted with varying severity on different countries in the region.

A scatter diagram similar to that in figure 2 was, therefore, constructed to explore the differential impacts of this drought (figure 4). The hypothesized country typology would imply that predominantly agrarian, rainfed economies would be expected to lie to the right of the 45 degree axis, because drought principally impacts the agricultural sector, while leaving other sectors largely unaffected. This is confirmed by the positioning of Malawi, Zambia, Mozambique, and Swaziland. More complex economies would be expected to be situated to the left of the 45 degree axis; drought would adversely impact not only agriculture but also on other sectors, as confirmed by the positioning of South Africa and Zimbabwe. South Africa is also located nearer to the (x,y) origin, as would be expected given its higher stage of economic development. In contrast, the performance of dualistic economies is indeterminate, depending on the performance of the extractive sector. In fact, Botswana and Namibia experienced strong growth in 1992, despite the weak performance of their agricultural sectors. Thus, the evidence for the 1991–92 southern African drought would broadly appear to confirm the validity of the country typology and "inverted U" hypothesis.

Further examination of the 1991–92 southern African drought also highlights an important temporal difference in the economic impact among the various economies. Recovery in rainfed agrarian economies largely depended on the return of good rains and the timely and adequate provision of agricultural inputs. However, recovery in more diversified developing economies was slower, requiring increased supplies of inputs to industries and a recovery in

Figure 4: Impact of the 1991/92 southern African drought on GDP and agricultural GDP - 1992 compared to 1991

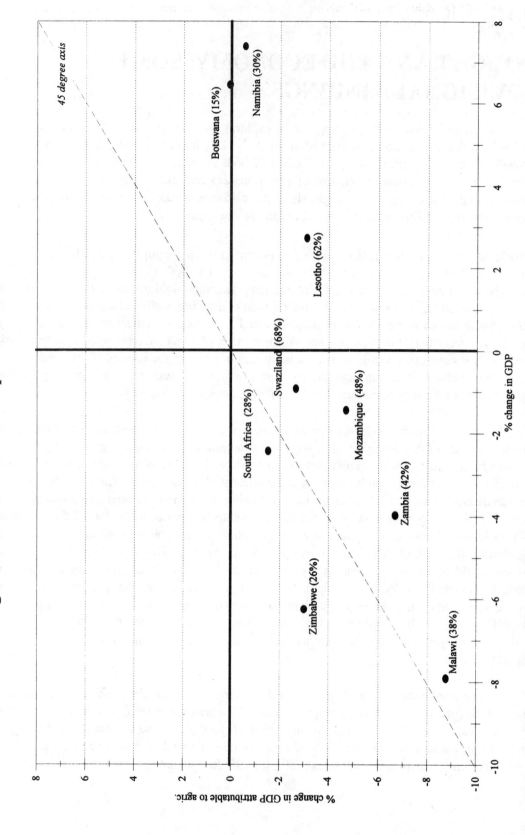

Note: 1992 maize yields as a percentage of 1989 yields are shown in parentheses as an indication of the severity of drought in each country.

Source: Based on data derived from the World Bank, 1995, "World Data 1995: World Bank Indicators on CD-ROM", Washington D.C. and FAO, various, "Production Year Book", Rome: Food and Agricultural Organisation

effective demand and credit markets. This difference is captured by considering changes in real GDP attributable to changes in agricultural GDP against changes in GDP for major southern African economies between 1991 and 1993 (figure 5). In both Zimbabwe and South Africa, the nonagricultural sectors continued to depress GDP in 1993, despite more rapid recovery of agricultural output after a favorable rainy season. For example, in Zimbabwe maize production rose to a five-year high in 1992–93, encouraged by incentive prices and a massive free input program for small farmers and good rains. In contrast, the country's nonagricultural sector remained depressed by lack of demand, high nominal interest rates, and tight credit conditions, in part a result of the drought (Benson 1997). Thus, although agricultural GDP grew by 48 percent in 1993, manufacturing output declined by 8.3 percent to its lowest level in real value terms since 1987, whereas overall GDP rose by only 0.9 percent. In the less complex economies of Malawi and Zambia, the agricultural sector accounted almost entirely for fluctuations in overall GDP. Namibia demonstrates how performance in the agricultural sector may have little impact on overall GDP in a dualistic economy with a large extractive sector.

Economic Sensitivity to Drought: Country Case Histories

The remainder of this chapter focuses on further empirical and qualitative evidence on the economic impacts of drought in the six case study countries indicated in chapter 1, with some additional references to Namibia's drought experiences.[24] Two of the countries considered, Burkina Faso and Ethiopia, can be stylized as simple economies. In both, as in many other SSA countries, the agricultural sector forms a particularly substantial part of GDP and the population is overwhelmingly rural. Kenya lies somewhere on the margin between a simple and an intermediate economy; it has an important diversified agricultural sector as well as a relatively large services sector. Almost two-thirds of microenterprises (including forestry and textiles) are directly based on agriculture (Block and Timmer 1994), but the manufacturing sector is still relatively small. Senegal, the one lower middle-income country considered, also has some characteristics of an intermediate economy; it has relatively strong intersectoral linkages reflecting past French colonial policy. Zimbabwe, which has one of the more developed manufacturing sectors in SSA, is the only economy that can be unambiguously classified as an intermediate economy. Zambia is essentially a dualistic, extractive economy. It has historically depended on copper production, particularly as a source of export earnings. Namibia's mining sector (primarily involving the extraction of diamonds, uranium, copper, lead, and zinc) also accounted for around a third of GDP in the 1980s as well as three-quarters of export earnings.

---

[24] The discontinuities resulting from conflict and recent independence, before which Namibia was closely integrated with South Africa, were the reasons for not undertaking a quantitative historical analysis.

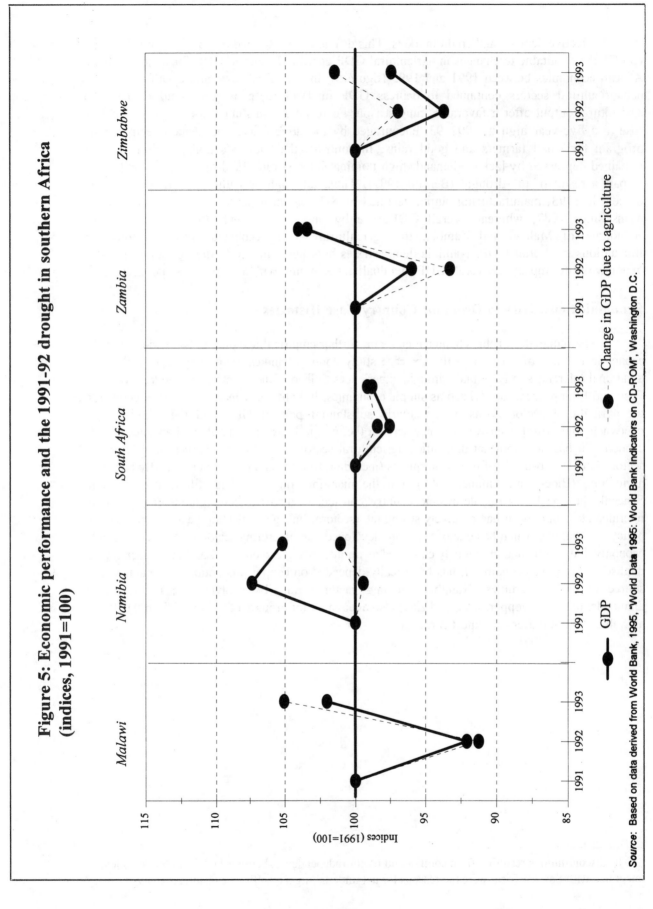

Figure 5: Economic performance and the 1991-92 drought in southern Africa (indices, 1991=100)

GDP

- - - Change in GDP due to agriculture

Source: Based on data derived from World Bank, 1995, "World Data 1995: World Bank Indicators on CD-ROM", Washington D.C.

Charting the economic impact of drought in the six case study countries in deviations in GDP and agricultural sector products from overall trends confirms significant differences in the vulnerability of the economies to drought (appendix figures 1–6).[25]

However, it also raised certain questions about other factors, beyond economic structure, influencing the outcome of drought events. The most striking correlation was indicated in the case of Zimbabwe, particularly since the early 1980s. In contrast, the Kenyan economy showed surprisingly little sensitivity to rainfall.

These relationships were explored more systematically using ordinary least squares regression techniques.[26] Reflecting the definition of drought given in chapter 2 and the sometimes considerable agroecological variability among countries, more sensitive country-specific drought indices than those used in appendix figures 1–6 were developed. These indices were based either on rainfall in the main food-growing regions of a particular country, where relevant data were readily available, or on the yields of a country's main food staples. The drought indices used together with the regression results are indicated in appendix table 1.[27] An asymmetric relationship between economic performance and rainfall was also assumed; above mean rainfall had no impact on the economy. This assumption was captured by constructing the drought indices as below period mean level indices. The analysis distinguished between the 1970s and 1980s to capture any changes over time in the drought sensitivity of economies.[28]

---

[25] Time trends were calculated for each country for the period shown in each graph. Deviations in GDP and agricultural products from these trends were plotted against deviations in national rainfall from a 30-year mean level. Varying time frames were used for different countries, depending on availability of data. For Zimbabwe, rainfall data were available on a monthly rather than annual basis. The rainfall index, therefore, lagged by six months, providing a more appropriate rainfall index because the main agricultural season extends from around November through to March and so spans two calendar years.

[26] The specification of the regression equations was highly simplified, ignoring other key factors determining economic performance such as the terms of trade or major political upheavals. However, fuller modeling of the determinants of economic performance was beyond the scope of this study.

[27] Difficulties in defining droughts are demonstrated by estimates of the severity of the 1983–84 drought in Ethiopia. One source indicates that, during 1983–85, 30 percent of farmlands at most were affected by drought (Kebbede 1992). Another source states that the drought "encompassed no less than 12 out of the country's 14 regions" (RRC 1985). A third, focusing on timing of rainfall, stated that in 1984 "the climatic conditions ranged from unfavorable to catastrophic throughout the country"; there were no dry season showers, first short rains failed, and the *belg* rains were both delayed and stopped early (World Bank 1985b). The period during which rainfall is measured is also important. For example, in Zambia annual rainfall exceeded the previous year's level in 1991–92 in most growing areas and did not constitute a "meteorological drought." However, the rains ended early, with low levels of rainfall in February 1992. This had a critical effect on maize production, which is most vulnerable to moisture stress in January and February (Riley 1993), and a national drought crisis was officially pronounced.

[28] This choice of periods is somewhat arbitrary and may not entirely reflect underlying economic trends. For example, in the case of Zambia declining copper export earnings from the mid-1970s had a major economic impact on the whole economy, most important, via the impact on export earnings. However, for the purposes of comparison among economies, it was decided to consider approximately common periods of analysis. In the case of Ethiopia, the 1974 revolution resulted in a major restructuring of the economy, implying that a comparison of the 1970s and 1980s would probably be invalidated by the nonneutral policy environment. The analysis, therefore, focused on the period 1974–75 to 1990–91 overall.

The results indicate that both GDP and agricultural GDP have displayed some sensitivity to downward fluctuations in rainfall, particularly during the more recent period analyzed and, in terms of countries, in the case of Burkina Faso.[29] However, droughts have had little apparent impact on manufacturing or industrial sectors, except in Senegal and Zimbabwe, a finding that is consistent with the four-country typology.

The Zambian results also support the hypothesis that a dualistic economy is partly cushioned from the impact of drought by its nonwater, intensive, extractive sector, with a relatively low, if significant, coefficient on the drought index as a determinant of GDP, despite the importance of rainfall as a determinant of agricultural GDP.[30] The one ambiguous case is Kenya, for which sectoral performance prior to the 1993 drought was only weakly correlated with rainfall, despite strong forward linkages between the agricultural and manufacturing sectors. Movements in international commodity prices and possibly the country's diverse agroecology appear to explain this finding (see below).[31]

A relatively speedy recovery from drought was indicated by a generally insignificant relationship between economic performance and lagged drought variables in the case study countries (Benson 1994). There is, however, some qualitative evidence that overall rates of recovery may be slower in relatively more complex economies, as already indicated.

In the long term, the impact of drought shocks is mitigated to the extent that drought-induced recessions are offset by higher growth during the upswing of the recovery period. However, droughts not only reduce sectoral output but also distort investment portfolios and restrict rates of capital formation, for example, with potential long-term consequences. Any such effects may be exacerbated if economic development objectives are set aside for the duration of the drought, weakening the links between relief and development as well as increasing the long-term costs of drought.

Other Factors Determining the Outcome of a Drought Shock

To understand more fully the factors determining the impact of a drought shock, documentation on the overall economic performance of the case study countries was examined. This review indicated a range of externally determined and domestic factors that can interact with

---

[29] T-statistics on the drought indicators were generally significant in the regressions against GDP and agricultural GDP, but F-tests indicated weak overall exploratory power in a number of cases, reflecting the reduced specification of the determinants of sectoral and GDP performance. However, there is some precedent for reporting results based on relatively weak adjusted R^2. For example, in exploring the effects of windfall gains, Collier and Gunning (1996) report results with some significant t-ratios but an adjusted R^2 of only 0.104.

[30] One of the most intriguing cases of dualism is South Africa, which is classified as upper middle-income in terms of per capita GDP and is the largest, most industrialized economy in Africa. The macroeconomic implications of the severe drought in 1991–92 were relatively modest despite maize import costs of some US$700 million (see box 2). However, with only 1 percent of white commercial farms accounting for 40 percent of agricultural output, there was a large "tail" of nonviable enterprises employing much migrant labor in drought-prone areas. Some 55 percent of the black population living in the former homelands depend on a combination of self-provisioning and remittances and are, therefore, food and health insecure in a drought.

[31] This apparent insensitivity could also reflect data problems, relating in particular to the underreporting of activities in the closely integrated sectors of subsistence agriculture and rural informal manufacturing.

rainfall variability to determine the extent and intensity of a drought shock, irrespective of the structure of a particular economy:

- **Prevailing economic conditions.** Perhaps most obviously, a weakened economy enhances the impact of drought, as demonstrated by the 1991–92 drought in southern Africa and the 1993 Kenyan crisis (Thomson 1995). The spread of the HIV/AIDS virus is also expected to heighten the economic fragility of a number of SSA economies, again exacerbating their vulnerability to drought and other exogenous shocks.[32]

- **International commodity price movements.** The six SSA economies that examined all depend on two or three export commodities for a significant part of their export earnings. Contemporaneous fluctuations in the prices of such commodities as well as of major imports such as oil can, therefore, play a major role in exacerbating or mitigating the impacts of a drought, particularly because trends in production in these countries, as in much of SSA, typically have little influence on world production.[33] Thus, the Zimbabwean economy was adversely affected by a weak international market for nonferrous metals as well as by drought in 1992–93. In contrast, improved groundnut prices together with strong growth in fish and fertilizer earnings counteracted the impact of Senegal's 1983–84 drought on its balance of payments. More generally, in Burkina Faso annual rainfall patterns and international cotton prices were negatively correlated during the 1980s; higher prices for the country's primary export partly offset the impact of lower rainfall. Movements in international commodity prices also partly explain the apparent insensitivity of the Kenyan economy to drought. During the severe drought year of 1984, when maize production declined by some 38 percent, agricultural sector production fell by only 3.5 percent. Tea and coffee production was little affected by the drought, and export earnings from these crops reached new all-time highs in 1984 and 1985, boosted by high world market prices and a rundown of domestically held stocks. Indeed, despite "massive" drought-related food imports, Kenya's foreign exchange holdings reached a three-year high of US$422 million in the second quarter of 1984 and twelve months later had fallen by only 8 percent (World Bank 1991b).

- **The structure of the agricultural sector.** Although agriculture is typically most directly sensitive to the level and timing of rainfall and dry spells relative to the agricultural cycle, the precise nature and extent of that vulnerability depend on a number of factors. These include the types and varieties of crop grown; planting techniques; the proportions of rainfed and irrigated production; the quality of cultivated land, including the scale of use of marginal lands; and the structure of land management and ownership. Government agricultural and food policies also play some role.

  For example, in the 1980s the Zambian government promoted the production of maize at the expense of more drought-resistant crops through various input and output subsidies

---

[32] For example, Ainsworth and Over (1992) report two studies by Cuddington (1991) and Cuddington and Hancock (1992), which estimated that the rate of growth of per capita GDP would be reduced by 0.1 percent in Tanzania and by 0.3 percent in Malawi under their most plausible set of assumptions about the AIDS epidemic.

[33] Kox (1990) estimated that, as recently as 1988, two commodities provided at least 60 percent of export earnings to thirty out of the forty-three SSA countries. Zimbabwe was the only country where the top two commodities provided less than 40 percent of total export earnings.

and other measures, resulting in a gradual increase in maize production to around 45 percent of agricultural GDP by 1993. However, commercial farmers began to diversify into other crops, while smallholder and emergent farmers' shares in maize acreage rose, increasing the drought vulnerability of national maize production to the extent that the latter groups of farmers cultivate a higher proportion of their crops under rainfed production and on more marginal lands. These farmers were also likely to have grown drought-resistant crops in the past. Meanwhile, the switch in production between the different categories of farmers was accompanied by the increasing cultivation of more moisture-sensitive hybrids. However, a more recent lifting of maize subsidies has seen a reemergence of other crops, including more drought-resistant sorghum and millet.[34]

In Zimbabwe, the relative shift in maize and cotton production from the large-scale commercial to the communal sector, the latter of which is heavily concentrated in lower potential marginal areas, has also been associated since 1980 with increased rainfall-related variability in agricultural production. During the period 1982–83 to 1992–93, regression analysis indicates that a rainfall level 10 percent below the 1969–93 national mean would be associated with a 25 percent reduction in maize yields from the communal sector compared with only a 17 percent drop in commercial sector yields. Meanwhile, a 30 percent reduction in rainfall would be associated with declines of 62 percent in communal and 47 percent in commercial sector maize yields.[35] These examples underscore the importance of a disaggregated approach in examining the sectoral impacts of drought and of taking into account the effects of structural change in assessing the drought vulnerability of both individual sectors and the wider economy.

- **Environmental degradation.** Increasing demographic pressures are resulting in the intensified utilization of more marginal lands in, for example, all the case study countries excepting Zambia.[36] These lands, by their very nature, are likely to be more vulnerable to adverse rainfall conditions. Furthermore, even less marginal lands are gradually losing productivity in a number of countries, such as Senegal, because of more intensive rainfed cultivation, which again increases the sensitivity of crop yields to weather conditions. Sahelian countries have also experienced increased aridity over recent decades, as already indicated. There is no consensus on the mechanisms underlying increasing degradation and, thus, no scientific basis for ascertaining whether or not this trend will continue (Hulme 1992 and 1996).[37] However, it does imply that indigenous agricultural and other

---

[34] This could also have important implications for levels of effective demand during periods of drought, because small farmers depended on maize to provide some 90 percent of their cash income as well as for domestic consumption during the early 1990s (Banda 1993).

[35] The results are sensitive to the choice of base and end years of analysis and to the rainfall indicator selected. The results reported, which are based on rainfall in critical winter months for selected stations in the various agroclimatic zones and which include a time variable in the regressions, explain 88 percent of variation in communal sector yields and 77 percent in commercial sector yields, with highly significant t-ratios for the rainfall variable.

[36] This may still be less an issue in Zambia because of its low rural population density and high level of urbanization for a relatively low-income country (Tiffen and Mulele 1993).

[37] The relationship between the process of environmental degradation and demographic pressures is the subject of much controversy. There are even some counterexamples to the demographic pressures hypothesis. For example, a study of Machakos District, Kenya, found that population growth was associated with a reversal of the process of environmental degradation in the medium to long term (Tiffen and others 1994). Benefits include increased food requirements, in turn, stimulating investments in land and technology; a larger labor supply; increased interaction of ideas, contributing to the development of

natural resource management practices, which evolved to suit historical rainfall patterns, may no longer be sustainable. Thus, vulnerability to drought may increase in the short term, before households and economies have fully adapted to the changing climatic conditions (Davies 1995).

- **Management of water resources.** The impact of drought on particular activities may be intensified if water resources have been poorly managed in the past, already lowering water "stocks" and implicitly accepting greater risk exposure. The experience of Zambia and Zimbabwe in 1992 demonstrates how the rules on the use of water may leave supply at risk to extreme events (see box 4). Various factors including demographic pressures, improved domestic access to water, rising standards of living, the expansion of irrigation networks, industrialization, and the growth of the tourism industry are currently increasing levels of water consumption across SSA, implying that the careful management of water resources will become increasingly important in determining the outcome of drought conditions in the future.

- **Cereal reserves.** Levels of food stocks held at the national and household levels are an important determinant of the short-term consequences of drought. They offer a timely response to impending food shortages and reduce short-term pressures on foreign exchange reserves. For example, maize stocks in excess of official minimum food security targets were important in ensuring adequate food availability in the aftermath of the 1984 and 1994 droughts in Zimbabwe. In contrast, much reduced stocks necessitated large-scale food imports to Zambia and Zimbabwe in 1992 ahead of international relief. However, large carryover stocks are also costly to maintain. Individual countries should, therefore, review acceptable levels of risk and the appropriate size of food reserves as well as the possibility of holding some reserves in a financial form instead (see chapter 6).

- **Migration and remittances.** Evidence from a number of SSA economies highlights the important role that inflows of remittances from migrant workers can play in reducing vulnerability to drought. These often involve migration to a neighboring country to work in the mining sector or to a coastal economy unaffected by drought, effectively spreading the benefits of drought-insensitive activities across borders. However, changing job opportunities also imply changes in the potential vulnerability of both labor-exporting areas and economies to the risk of drought. In the case of Burkina Faso, for example, there was a marked decline in external workers' remittances in the early 1990s, owing to worsening economic conditions in neighboring countries. Remittances represented an estimated 31 percent of total export and private transfer earnings in 1990–94, compared with 40 percent in 1980–89, raising concerns about the country's future vulnerability to drought.

- **Internal or external conflict.** A conflict will precipitate government expenditure in domestic war-related industries and services, effectively maintaining some level of

---

new technologies; and economies of scale in the provision of social and physical infrastructure, for example, increasing access to markets and raising standards of living. However, it has yet to be demonstrated whether these phenomena have or potentially could occur in other regions of Sub-Saharan Africa.

economic activity even during periods of drought. Conflicts can also localize the impacts of drought to immediately affected regions by disrupting the flow of goods and services among regions. Such effects have been demonstrated in the case of drought in Ethiopia, as already noted. Moreover, conflict may also reinforce the impact of drought to the extent that it disrupts agricultural and other productive activities and renders affected populations more vulnerable to the impact of events that in a more secure environment were within the range of normal coping practices.

Box 4: Lake Kariba and the 1991–92 Drought

Zambia and Zimbabwe rely on hydroelectric power generation for a major part of their electricity supply. During and in the aftermath of the 1991–92 drought, both countries experienced serious electricity shortages and also faced major threats to the continued supply of urban drinking water. However, the difficulties reflected less the impacts of the 1991–92 drought than the long-term mismanagement of common water resources. The drought simply proved to be the final trigger.

In the 1960s, climatologists had identified a statistically robust 18-year cycle in the summer rainfall region of southeastern Africa and successfully forecast the wetter and drier periods of the 1970s and 1980s respectively (Tyson and Dyer 1978). However, despite the predicted sequence of years of lower rainfall in the Kariba and Kafue catchments, during the 1980s the Zambian and Zimbabwean electricity-generating authorities had continued to base levels of water offtake for power generation at Kariba on the average intake for the relatively more favorable 1970s. Thus, offtake exceeded the rate of inflow into the lake by an average of 16 percent during the 1980s, making the system increasingly vulnerable to further rainfall anomalies, as was clearly demonstrated in the aftermath of the 1991–92 drought. The electricity curtailments in Zimbabwe alone were estimated to result in a Z\$560 million (US\$102 million at the 1992 rate of exchange) loss in GDP, a Z\$200 million (US\$36 million) loss in export earnings and the loss of 3,000 jobs. Such reckless management may have partly reflected pressures to minimize short-run costs of power generation in the face of the large operating deficits of the Zimbabwe Electricity Supply Authority.

The Policy Environment

The existing policy framework can also play a fundamental role in determining the impact of drought as well as other exogenous shocks, as already indicated in chapter 2. In countries where large areas of the formal economy are highly regulated, thus, effectively constraining levels of investment or imports, for example, the impact of a drought shock is minimized by existing binding constraints, as in Zimbabwe in the mid-1980s. The Zimbabwe government's trade policy also inadvertently influenced the impact of the 1991–92 drought, because uncertainty about the continuation of a more liberal import regime and speculation created by the unstable exchange rate had resulted in a gradual buildup in stocks of imported inputs in 1990 and 1991, so that producers entered the drought period with considerable inputs already in stock.

Conversely, government policies can exacerbate the impacts of drought. In Ethiopia, for example, the impacts of the 1982–84 drought were compounded by the Derg government's simultaneous implementation of a socialist system of collective ownership and centralized direction, which effectively undermined the ability of rural households to cope with the impacts of the drought. Namibia's experience in 1992 also highlights how a drought can in turn force other issues—in this case, chronic poverty—onto the government policy agenda, both increasing the costs of relief and entailing long-term implications for levels of government expenditure (Thomson 1994, Devereux and others 1995, Namibia 1997).

Governments can also use policy actions to influence the nature and scale of impact of drought events in a more deliberate fashion. Indeed, droughts pose certain policy choices, particularly relating to public finance and the external sector. These policy-related dimensions of a drought shock are illustrated in the next section, including evidence from Zimbabwe and Namibia on the budgetary implications of drought.

Drought and the Budget

Droughts have potentially important implications for government policy, first and foremost via their impact on the budgetary balance. A drought shock is likely to reduce tax revenue via a decline in income, employment, and exports. The revenue of utilities will also be adversely affected by income impacts of the shock-induced recession on effective demand and increased nonpayment. Revenue from parastatal utilities in the water and hydroelectric power generation sectors is also likely to fall, owing to some combination of lower output, effects of recession on demand, and simultaneous upward pressures on costs from the crisis provision of supply. Other parastatals could be similarly affected by the recessionary impact of drought.[38]

On the expenditure side, governments may be confronted by increased expenditure on relief, social welfare, health and water supplies, consumption-related subsidies on food distribution, and the logistical costs of drought-related imports. Law and order services could also be put under greater pressure by a rise in crime, in turn associated with temporary unemployment, migration, and increased destitution. In addition, there are likely to be pressures for the increased provision of subsidies and credit to affected productive sectors, including public utilities, both because of the direct impact of water shortages on their operations and because of reduced demand. For example, the Zimbabwean government had to meet the increased losses of the Agricultural Financial Corporation, the Grain Marketing Board (GMB), and the National Railways of Zimbabwe as a direct consequence of the 1991–92 drought. Prior to the deliberate rundown of maize stocks in 1990–91, the GMB had typically imposed heavier burdens on government finances during years of surplus than of years of deficit maize production because of particularly high storage and disposal costs, again demonstrating the importance of the policy environment in determining the impact of a drought.

---

[38] In the case of electricity generation, more costly thermal power plants may have to be used to provide a higher proportion of the load because of restrictions on hydroelectric supply. Load shedding could result in a reduction in revenue.

Figure 6: The impact of drought on the growth rates of total debt stocks

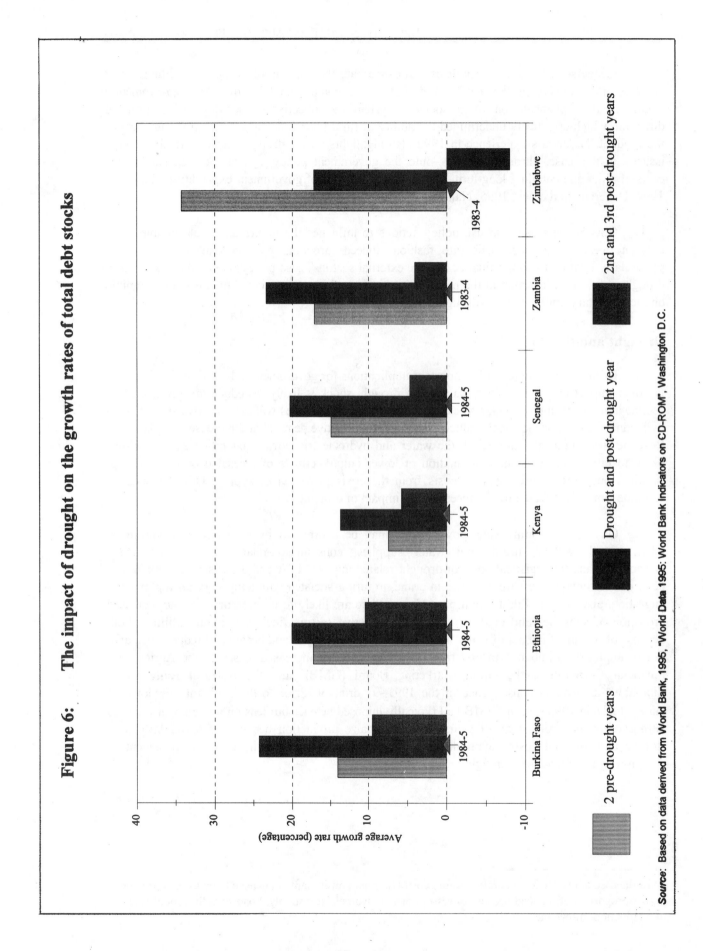

Source: Based on data derived from World Bank, 1995, "World Data 1995: World Bank Indicators on CD-ROM", Washington D.C.

Increased budgetary pressures, resulting from lower revenues and higher expenditure, can be met either by raising additional financing or by reallocating planned government expenditure or by a combination of the two. There are three basic potential sources of additional finance: borrowing, higher taxes, and increased charges for publicly provided goods and services.

Government can increase borrowings from domestic and external official and private sources, although its ability to do so is in part determined by its existing level of indebtedness, its relationship with official lenders, and the confidence of private lenders in the economy. Indeed, there is evidence of increases in both external and internal borrowing during periods of drought. The former is indicated in figure 6, which shows increases in debt stocks in five of the six case study countries in the early 1990s as a consequence of drought. The exception was Zimbabwe, whose debt fell as a result of a deliberate long-term policy of debt reduction. Increased indebtedness has obvious implications for future levels of debt servicing and thus for the availability of foreign exchange to finance capital imports. For example, in Ethiopia, forecast debt interest payments alone, excluding amortization, for 1985 increased by 9.2 percent during the drought year of 1984. Meanwhile, a rise in internal borrowing is illustrated by the actions of the Zimbabwean government in 1992. As noted earlier, the government offset the implicit monetary expansion entailed in this increase in borrowing by maintaining the existing high levels of interest. This in turn contributed to an annual 5.2 percent fall in private sector investment as compared with the previous year, thus delaying the expected refurbishment of domestic industry at a time when barriers to external competition were removed.

Higher taxation is not unproblematic as a source of additional government financing because of the direct and recessionary effects of the shock. In Zimbabwe, for example, drought levies were imposed on company taxes in both 1984–85 and 1987. However, the rate of company taxation was reduced to help alleviate financial difficulties that had arisen partly as a consequence of the drought. Higher electricity or water prices, despite supplies being temporarily less satisfactory as well as recessionary pressures, will be contentious but may be effective where demand is inelastic. Attempting to reduce subsidies for food and fertilizer or in the social sectors also involves challenging decisions in a crisis.

Reallocation of planned government expenditure may occur within or between sectors and between capital and recurrent spending with varying opportunity costs. In both Zimbabwe and Namibia in 1992, for instance, drought-related expenditures apparently led to a shift in the composition of investment programs, delaying some nondrought-related projects. In Zimbabwe, several nonwater-related projects planned under the Second Five Year National Development Plan (1991–95) were delayed, whereas the implementation of some planned water-related projects was brought forward and further new ones begun. Additional government finance was transferred from capital to recurrent expenditure (Benson 1997). Similarly, in Namibia it was announced that the initial funding of the drought relief program would be met by postponing the upgrading of the trans-Caprivi highway, although it was actually unclear precisely where the funds for the drought relief program originated (Thomson 1994). Further research would be required to analyze the precise long-term economic costs of such reallocations of government expenditure, but they almost certainly reduce rates of growth and have immediate social costs. For example, an estimated 58 percent of the Namibian Drought Relief Program expenditure involved the provision of goods, services, and financial transfers, which would not have been provided in the absence of drought. These expenditures had no long-term benefit although some elements of the program—for example, in the area of water supply—fitted in with existing work and priorities (Thomson 1994).

There is also some evidence that levels of government staffing may not be increased sufficiently to deal with the extra drought-related workload, thus, displacing normal activities and effectively entailing an additional, although intangible, switch of government expenditure. For example, in Zimbabwe, in 1992, funding was switched from preventive health care, such as immunization programs, to emergency activities (Tobaiwa 1993).

Consideration of Drought Shocks in Economic and Development Planning

Government and donor policy and planning documents were also examined to assess the extent to which drought risks are recognized as a potential economic threat and taken into account in economic strategy and policy design.

Box 5: The Role of Agricultural Growth Multipliers

The size of agricultural and nonagricultural multipliers plays an important role in determining the second round and subsequent effects of drought shocks.

Recent studies have found that agricultural growth multipliers in Sub-Saharan Africa may be much higher than previously calculated, possibly exceeding those of nonagricultural multipliers. In the case of Kenya, for example, Block and Timmer (1994) estimated that over a four-year period an increase of Ksh100 million in agricultural value added would increase nonagricultural value added by just more than Ksh56 million and agricultural value added (via further investments in agriculture) by approximately Ksh8 million, implying an agricultural multiplier of 1.64. In contrast, they estimated a much lower nonagricultural multiplier of 1.23. In other words, according to these findings, a given increase in agricultural income will have a much greater impact on national income than an equivalent increase in nonagricultural income. In the event of a drought, these linkages will obviously work in reverse; declines in the agricultural sector will impact severely on other sectors.

Similarly, Delgado and others (1994) have estimated high agricultural multipliers for other SSA countries, including values of 2.75 in Burkina Faso, 2.48 in Zambia, and 1.97 in Senegal. These values are slightly higher than previous estimates, reflecting a decision to include food grains as nontradeables, as is considered more appropriate in an SSA context.

Further evidence could be provided in support of the hypothesized increase in vulnerability to droughts in the early stages of development by estimating agricultural and nonagricultural growth multipliers for countries at differing stages of development. If correct, the "inverted U" hypothesis would imply that agricultural multipliers increase in the early stages of development before declining at a later stage.

In some southern African countries, policymakers finally appear to be displaying a greater awareness of the economywide threat of drought. For example, the Zimbabwe government in its appeal for drought assistance in 1995 stated that "drought is now a permanent feature in Zimbabwe. It is, therefore, necessary that Government makes long-term strategies to deal with this phenomenon" (Zimbabwe 1995: 9). It then went on to list critical areas requiring action: promotion of appropriate economic activities in specific agroecological areas, increased production of small grains where these are the staple crop, construction of large-scale dams, establishment of a Permanent Drought Mitigation

Unit to plan and implement long-term drought mitigation and recovery measures, and provision of adequate rations for those affected by droughts. However, even this remains a rather narrow perception of the impacts of drought, and further moves to push drought onto the broader development agenda are required. Regionally, SADC is now emphasizing the need to take the wider economic costs of drought into account in sectoral preparedness and training.

For some SSA countries, development policy analysts are advocating increased emphasis on the agricultural sector as a source of growth. The implications for drought vulnerability are not clear-cut. In Kenya and Zambia, for example, agriculture is being promoted as a primary source of growth and increased export earnings, because the scope for further short- to medium-term economic growth in other sectors, particularly public services, diminishes, and pressure to secure long-term sustainable growth increases. However, food staples must not be neglected as governments and donors increasingly emphasize the development of other crops. Instead, considerable efforts should be made to reduce fluctuations in food staples, including subsistence and livestock production. Furthermore, the risk of drought should be properly incorporated into all agricultural plans. Work by Block and Timmer (1994) and others emphasizes the potential pitfalls as well as the potential economywide benefits of higher agricultural growth (see box 5 above). The challenge is to reduce vulnerability to drought.

CHAPTER 5:

DROUGHT AND STRUCTURAL ADJUSTMENT PROGRAMS

Droughts and other exogenous shocks affect both the performance of economic aggregates and underlying behavioral relationships, implying that structural adjustment programs (SAPs) applied during periods of good rainfall may not be appropriate during droughts.[39] However, the relationship between the process of structural adjustment and economic resilience to drought is complex and raises a number of issues. These can be organized under six broad lines of inquiry:

- Is drought a contributing factor to the adoption of a SAP?

- Does drought impede the progress of reform programs?

- Does drought exacerbate the potential adverse short-term impacts of reform?

- Is drought a factor contributing to the abandonment of reform programs?

- Does the existence of a SAP influence the nature and volume of drought relief measures?

- Does successful structural adjustment reduce the vulnerability of an economy to drought?

Adoption of Reform Programs

There is some evidence that natural disasters may be an important short-term trigger contributing to the adoption of a structural adjustment program. For example, Killick and Malik (1992: 604) found in a survey of seventeen randomly selected developing countries with IMF stabilization programs that, in six cases, natural disasters—and in four, drought specifically—had been an "important, perhaps dominant factor" in the decision to adopt a program. Among the six case study countries, Senegal provides a further example of the role of drought in leading to the adoption of structural reform programs. In 1980 the Senegalese government adopted a stabilization program, financially supported by the World Bank and the IMF, in response to mounting long-term economic problems arising as a consequence of growing fiscal and budgetary deficits. The severe 1979 drought, as well as an expansion of the public sector and the 1979 oil price shock, was identified as having contributed to these economic difficulties.

However, whatever the factors contributing to the adoption of a reform program, efforts to launch a major program may in fact be inappropriate in such circumstances. Experience has demonstrated that structural adjustment needs to occur against a backdrop of macroeconomic

---

[39] Stewart (1993: 358) also notes that "no effort has been made to consider what adjustment policies are appropriate during war." As in the case of drought, normal policy packages are not appropriate, in part because of the "critical need" to maintain government expenditure and thus normal economic and social functions as well as to finance reconstruction and social relief functions at the same time as having high levels of military expenditure. This is in direct conflict with the expenditure-cutting tendency of most adjustment packages.

stabilization, a situation that does not obtain during periods of major drought. Indeed, an internal World Bank report concluded that certain countries, including those recovering from natural disaster or war, were "plainly inappropriate" for adjustment. In these countries, emergency or bridging operations are considered more appropriate.

Progress of Reform

SAPs do not typically anticipate droughts in their design, despite the fact that droughts can have major implications for the progress and ultimate success of reform programs. The progress of a reform program can be considered from three perspectives:

- The pace of implementation of the various reform measures

- The degree to which private economic agents respond to the reform measures as expected

- The extent to which economic targets are achieved.

The stage reached in an adjustment program at the time the drought shock occurs is a critical factor in all these regards. The impact of a drought shock would be expected to be less significant during the later stages of a reform program. The pace of implementation of reforms in the event of a drought shock is also partly determined by the resolve and political strength of the reforming government, particularly during the earlier phases, and the level of donor resources potentially at stake if the reform program is thrown off course.

The recent southern African drought offers some evidence on the ways in which a drought shock interacts with the progress of reform programs. In the case of Zambia, the 1991–92 drought had an adverse effect on the expected response of the private sector to the reform program, including to the liberalization of product and factor markets and the dismantling of price controls, and on the achievement of the original economic targets. Inflationary concerns partly related to drought also led to the adoption of a cash budget at the end of 1992. This unfortunately reverberated in the recovery of the maize sector because the government reduced its maize marketing operations as part of its cost-cutting measures, whereas high domestic interest rates severely hindered the growth of private market operators. As a consequence, a substantial quantity of maize was left unpurchased in 1993, in turn creating further hardships for farmers in addition to those inflicted by the previous year's drought. The government was consequently obliged to issue promissory notes to farmers for the purchase and delivery of the crop during the 1993–94 marketing year ("Adjustment in Africa" 1994).

In Zimbabwe, the 1991–92 drought also adversely affected the reform process, hampering efforts to reduce the budget deficit and restructure the civil service and parastatals. Intensified budgetary difficulties in turn frustrated government efforts to reduce domestic borrowing and so partly curtailed the expected domestic supply-side response to the reform program, which was critical to its success. However, the drought shock also partly speeded up the restructuring of the manufacturing sector, because increased liquidity constraints forced enterprises to carry smaller stocks, shed surplus labor, and adopt less costly, efficiency-enhancing methods of production.

Progress of reform in the agricultural sector is perhaps most susceptible to fluctuations in weather conditions. Drought shocks rekindle overriding concerns about food security, weakening governments' resolve to proceed with reforms and reducing the willingness of the private sector to become involved in market reforms. In Kenya, drought has clearly had a considerable effect on the pace of cereals market reform; a discernible pattern of reluctant agreement to reform has been followed by retraction in the face of disruption to domestic cereal markets as a result of drought (Thomson 1995). However, it is less clear how this frustrating process of advance and retreat could have been avoided and how the current liberalization process can be kept on track. Some way has to be found to strengthen the constituency for liberalization while satisfying other powerful interests, particularly, the producer lobby. This is made more difficult by the fact that the main opposition to government intervention in Kenya is based on its financial costs.

Short-Term Adverse Impacts of Reform

Droughts probably exacerbate the short-term adverse impacts of reform first and foremost via their effect on vulnerable groups. Droughts typically reinforce the inflationary impact of the removal of subsidies on basic food and other commodities, to the extent that governments continue this policy, and also contribute to job losses. Droughts can also intensify the short-term adverse economic impacts of reform more generally, for example, by prolonging tight monetary policy with implications for the rates of investment, as occurred in Zimbabwe in 1992–93.

Abandonment of Reform Programs

Senegal's experience in 1980 provides the only example from the case studies of a country where drought contributed to the abandonment of a reform program. This instance is particularly ironic because, as already indicated, drought was also a factor contributing to the decision to adopt the reform program in the first place. More recently, the international financial institutions (IFIs) appear to be more willing to modify structural reform programs in the aftermath of drought, as indicated by experience in both Zambia and Zimbabwe. This flexibility possibly reflects greater experience in the implementation of reform programs.

Governments also appear to have come under growing pressure to sustain reform programs during periods of economic difficulty resulting from a drought or some other exogenous shock. This is because international assistance has become increasingly conditional on domestic policy reforms; even bilateral donors often link the continued provision of assistance to compliance with the conditions of IMF and World Bank agreements. Governments are also sensitive to domestic political constituencies such as producer groups and urban consumers. During periods of drought, these potentially conflicting constraints require a careful balance of policies to avoid in a country, such as Kenya, a frustrating process of advance and retreat in the process of liberalization (Thomson 1995).

Responses to Drought

The existence of reform programs and more generally relationships with donors, particularly the IFIs, seem to play an important role in determining the nature and level of the international response to drought. This partly reflects a growing reluctance to provide aid to nonreforming countries, particularly those from which the IFIs have actually withdrawn assistance. In contrast, during periods of drought, donors may provide additional assistance to reforming governments to prevent the drought shock derailing the reform process. In the case of Zambia, some senior officials are firmly of the opinion that, if a drought of a similar magnitude to that of 1991–92 had occurred in 1987 when the government had just broken off discussions with the IMF, the country would have received little external assistance.[40] In contrast, the sizable level of international assistance provided in response to the 1991–92 drought reflected serious concern that Zambia's democratic regime, which was being held up as a model for other countries in Africa, should not be shaken and that the adjustment program approved by the IMF and the World Bank "should in no way be jeopardized by the drought" (Seshamani 1993: 5). Zimbabwe's experience in 1992 supports this view of the likely international response to a drought shock; several donors have conceded that they were supporting drought relief partly to keep the SAP on course (Thompson 1993).

An ongoing SAP also implies that an economy's performance is already being closely monitored, thus, providing early and credible indications of economic difficulties emerging as a consequence of drought. This, too, may influence the nature and magnitude of the international response to drought, in terms of the availability of financial support to meet additional balance of payments pressures as well as of more conventional emergency food aid and other relief measures. For example, this was the case with the combination of measures to ease balance of payments pressures on Zambia and Zimbabwe in 1992, which included additional untied financial support and assistance tied to drought-related imports. The apparently unprecedented action of the IFIs in monitoring and addressing the financial implications of the drought-related imports of all the affected countries of southern Africa is an important and positive precedent for future international responses to crises in SSA.

Some of the financial difficulties encountered in drought management have been attributed to the excessively inflexible pursuit of adjustment targets during a crisis. In 1992–93, in Zambia, the tight public expenditure policy being implemented as part of the adjustment process restricted the government's ability to raise external financing for drought-related measures because of the lack of counterpart Kwacha resources. As a result, international funds were not fully utilized. Limited local currency funds also created difficulties in meeting internal transport, storage, and handling costs, thus, hampering the transport of relief supplies. The gradual depreciation of the Kwacha over the period of the relief operation as the exchange rate was liberalized also created problems in fixing transport rates, resulting in some delays in delivery of relief supplies (Seshamani 1993). In the case of Zimbabwe, the Ministry of Finance, in seeking to maintain its relatively tight controls on expenditure, may have held up disbursement of additional external assistance linked to the drought (Thompson 1993). The balance of causation—rigid pursuit of targets or inefficiency in public expenditure management—is always debated in individual cases. Nevertheless, such concerns indicate the need in a crisis situation for careful review of expenditure priorities, fiscal targets, and efficiency in realizing revenue, for example, from counterpart funds.

Structural Adjustment and Drought Vulnerability

The World Bank defines structural adjustment as "reforms of policies and institutions [that] improve resource allocation and increase resilience to future shocks" (World Bank 1988). In the long term, the successful implementation of a structural reform program should reduce the overall vulnerability of an economy to drought shocks by restoring economic growth. Particular aspects of successful structural reform programs, such as reduced government budgetary deficits, higher levels of domestic savings, liberalization of external trade controls, devaluation of currencies to competitive levels, and elimination of pricing and market distortions could also play a more direct role in mitigating

---

[40] Three adjustment credits were approved for Zambia in 1985–86. Negotiations between the IMF and the Zambian government were suspended in 1987 after Zambia abandoned its reform program and ceased debt-servicing payments. Adjustment lending was resumed in 1991 after the country had cleared its debt arrears.

the impact of a drought. For example, a number of SSA countries have already implemented changes in agricultural pricing policies and lifted some regulations on agricultural marketing, prices, and imports. Private traders and millers are now playing an increasing role in domestic markets. This could potentially reduce vulnerability to drought by permitting early market responses to droughts, which avoid substantial price increases and the need for large food aid imports. However, elimination of agricultural incentives could also stimulate shifts in cropping patterns toward ones more in keeping with climatological conditions, as has recently been seen in Zambia, for example.

A recent World Bank report concluded, however, that no African country had yet successfully completed the reform process, that there were grounds for concern that reforms undertaken to date were "fragile," and that adjustment alone was not sufficient to "put countries on a sustained, poverty-reducing growth path" (World Bank 1994: 2). A SAP is only one element of broader government plans and policies. If the latter do not incorporate proper management of natural and other resources and of drought risks, then SAPs alone can do little to reduce the vulnerability of an economy to drought. Recognition of drought as a fundamental obstacle to economic development, particularly in simple agrarian economies, must necessarily be part of more general development and economic strategies if they are to succeed.

In conclusion, there are considerable potential tensions between satisfactory implementation of SAPs and postdrought recovery programs. Donors should also be sensitive to the particular difficulties posed by droughts, including heightened political pressures. This requires careful monitoring and perhaps the modification of SAPs.

CHAPTER 6:

SUPPORTING DROUGHT CRISIS MANAGEMENT AND MITIGATION MEASURES

This study has shown how economic structures and resource endowments interact in determining the impacts of drought shocks. The typology developed to focus on these differences is used here as a framework for examining the policies and instruments available to the IFIs and bilateral donor agencies in supporting drought crisis management and mitigation measures in SSA.

In considering issues of donor policy, the additional dimension of *governance* should also be taken into account. In the context of drought, good governance implies a government genuinely committed to the well-being of all communities and regions. It will accord the highest priority to responding to the social and economic threat posed by drought and to preventing the occurrence of such disaster. The capacity to mitigate the effects of disasters is also implied, a capacity that in some other circumstances, especially those of conflict, is severely curtailed. The interaction of these three factors—economic structures, resource endowment, and governance—provides a framework for distinguishing among broad but distinctive drought response and mitigation strategies.

SAPs are being implemented in many SSA economies. This is, to borrow Sir Hans Singer's phrase for describing food aid policy, both a challenge and an opportunity. Structural adjustment involves a highly constrained policy framework of objectives and targets within which a government is expected to respond to a crisis. However, an SAP implies that donors, together with government, are monitoring the economic and financial situation closely. The presence of a SAP may also imply that external resources are potentially more readily available to counteract the negative economic consequences of a drought shock.

Appropriate Aid Instruments for Drought Crisis Management

The selection of available instruments has to be sensitive to the structure and resource endowments of the recipient economy, the current economic policy environment, and broader questions of *governance*. This latter point may be made by contrasting, for example, the situation in 1992 in Zambia and Zimbabwe, which have relatively effective governments with a relatively broad mandate, with that of Malawi and Mozambique, which less effective governments and large populations displaced by conflict. Alternatively, the situation in Kenya in 1979–80 or again in 1984 was from the perspective of governance and the confidence that the donors had in government different from that in 1993. Similarly, in considering dualistic mineral economies, the situation in Botswana and Namibia has been different from that in Niger, for example. Therefore, the selection of instruments will unavoidably take into account an assessment of the capacity and commitment of government to drought crisis management.

The instruments available to the donor include *financial aid* for balance of payments and budgetary support, *program food aid* for similar purposes, *bilateral project aid, emergency financial aid,* and *emergency food aid. Advocacy* is another policy instrument that IFIs and bilateral donors as well as NGOs employ to influence the actions of other important actors. After a careful review of the evaluative evidence on responses to drought crises in the early 1990s, particularly the more complex, better integrated economies of southern Africa, and taking into account the typology and instruments available, a broad hierarchy of instruments in terms of their effectiveness is indicated.

Intermediate, More Complex Economies

In thinking about future policies, the more complex SSA economies provide a useful starting point for considering how the international community or individual donor can support drought management and mitigation. The process of development is gradually increasing the complexity within most SSA economies in terms of international linkages and commercial and financial flows. Political democratization, most obviously in southern Africa, and economic liberalization are accelerating the process of transformation to more complex systems of financial flows, integrating markets within countries and regionally. This process involves countries as diverse as Malawi and Mozambique or Eritrea and Ethiopia. It has been tentatively suggested above that this transformation initially exposes an economy to drought shocks rather than reducing its vulnerability and that the challenge of drought management includes but is much more than that of assuring food and potable water for affected people and their livestock.

In a more complex economy, the shock is likely to be widely diffused throughout. The urban industrial economy will be affected both directly and indirectly via intersectoral linkages and expenditure multiplier effects. An extreme drought is likely to precipitate a wide, more generalized economic recession in which the overall loss in GDP substantially exceeds the direct effects on agricultural production. With a more complex financial system, remittances to rural areas may help some of those within affected groups to cope with the impacts. But because migrants are predominantly employed as manual laborers (men) or in low-paid service activities (women), this also spreads the effects of the drought more widely through the low-income groups in society.

In terms of *governance*, there will be considerable social pressures to intervene to help affected populations, to limit the effects on urban consumers, and to prevent the drought and the associated recession from having severe economywide impacts. There will be strong pressure on domestic finances from both the revenue and the expenditure sides. There are likely to be foreign exchange pressures because of a direct reduction in export revenues and increased food imports.

Financial Aid for Balance of Payments and Budgetary Support

The provision at low cost of *additional financial aid* to provide balance-of-payments and budgetary support should have the highest priority in such economies. This aid should be focused on meeting the direct costs of the drought response but in a way that counteracts the recessionary effects of the economic shock. It should also be rapidly disbursed to minimize the negative impacts of the drought shock on government foreign exchange reserves as a consequence of additional food imports and reduced export revenues. Rapid disbursement is also important in minimizing additional internal funding requirements. In the context of a SAP, flexibility in the use of funds already programmed is important to avoid a drop in disbursement. Flexibility in terms of modifying SAP targets may also be required, because governments will face difficulties in staying within financial limits agreed on prior to the crisis in the face of the additional costs of drought relief and losses of revenue.

Cost of Financial Assistance

Efforts should be made to minimize increased medium- and long-term indebtedness. For example, the IMF has a cereal import component within its Compensatory and Contingency Financing Facility (CCFF). In practice, fresh arrangements under the CCFF were not made by any of the case study countries for recent drought-related cereal imports or, indeed, by other countries affected by the 1991–92 southern African drought. This reflects the relatively high-cost terms of the CCFF and the problems of concluding a rapid agreement in response to immediate financial requirements. Instead, affected countries have preferred to seek bilateral grants and more concessional IDA credits in the form of Emergency Recovery Loans.

Program Food Aid for Sale on Local Markets to Generate Counterpart Funds

If it can be rapidly committed and delivered, program food aid is almost as effective as financial assistance. The main questions are whether it will be timely, cost-effective, or appropriate in terms of providing the type of food that can be readily absorbed in the market of the recipient country. Where internal budgetary support is intended, counterpart funds arrangements may require special attention. These should be appropriate to allow rapid disbursement to meet crisis-related expenditure, for example, on the internal transport costs of relief, and not just concerned with satisfying donor accountability requirements. In Zambia and Kenya in 1993, the long lead times from initial request to actual arrival were such that some of the food aid became part of the problem of postcrisis food system management (Hannover and others 1996, Legal and others 1996). Support for commercial imports by public agencies or the private sector may be more appropriate, timely, and cost-effective (Clay and others 1996).

Bilateral Project Aid Including Technical Cooperation and Funding of Local Costs and Import Components

Again, it is essential to sustain the flow of project funds, thereby maintaining activities throughout the drought shock. There is also some scope for "quick action" projects or accelerated implementation. Unfortunately, the design of projects does not normally take account of the risk of a serious drought; this has been true even of many agricultural projects in drought-vulnerable areas (World Bank 1991c). The problem of sustaining project disbursements is probably also linked to lack of local funds where drought relief programs are crowding out other expenditure (e.g., Thomson 1994).

Emergency Financial Aid and Emergency Food Aid

Financial aid variously labeled as disaster, emergency, or relief assistance is provided for the direct relief of affected populations through NGOs, the United Nations, and possibly government agencies. In practice, the aid usually supplies specified goods, such as seeds, medicines, and trucks (often tied for bilateral donors) or specific emergency measures (such as repair of the water supply and sinking bore holes). Emergency food aid is provided for direct relief, again usually indirectly through NGOs or the United Nations. Where there is good governance, there is a lower priority for either emergency food or emergency financial aid, both of which are relatively inflexible instruments because they are tied to direct relief. There will doubtless be scope for NGO activity and special U.N. programs. However, emergency aid is costly to deliver and raises all sorts of targeting issues. Nor does such direct aid address the economywide aspects of the shock. Inflexibility, especially of emergency food aid, is another limitation of such aid, which usually cannot be redirected to reconstruction or more general economic support if the end of the drought and a rapid agricultural recovery reduce the need for direct distribution programs. Zambia, Zimbabwe, and Malawi all experienced such absorption problems; relief food aid arrived at the end of the drought in 1993 (Callihan and others 1994, Clay and others 1995, World Bank 1995c).

Simple and Conflict-Affected Economies

The effects of a drought shock in a simple, less complex economy will be concentrated in the rural sector. The impacts are likely to be particularly severe on those involved in self-provisioning and in marginal environments. Outside the rural economy, the direct impacts may be relatively limited. Impacts on the balance of payments from reduced agricultural production will depend on resource endowments and import requirements for drought crisis management. In terms of governance, there is often a lack of awareness of the severity of the crisis because those in the capital city are insulated by subsidies and food imports from the early impacts of the drought. At least in the first instance, government may choose not to take any action. Indeed, unless there is *good governance* that is sensitive to the impacts of the shock on the rural economy, there is a real danger of a drought-famine syndrome. Examples include some Sahelian countries in 1969–74 and again in 1982–84 (Glantz 1987). However, the effects of the drought on the balance of payments and the financing of additional drought response measures will put pressures on government to respond.

Countries affected by conflict become economically less complex: they are effectively experiencing a process of underdevelopment. Infrastructure and productive capital are destroyed or suffer from lack of maintenance. Markets become less integrated. Potentially vulnerable areas and groups become less resilient with lack of tools and inputs and depredations by foraging combatants. Impacts of shocks are, therefore, intensified within the immediately affected area because they cannot be more widely diffused. This was the situation in Ethiopia and Eritrea prior to the collapse of the Derg, in Mozambique up to the Peace Accord, and still in drought-prone Somalia and Sudan (Webb and others 1992, Macrae and Zwi 1994, Jean 1993).

The impacts of drought in both simple and conflict-affected economies are likely to be concentrated in the rural sector and have less general recessionary impact. Consequently, more targeted interventions are needed, for example, in the form of food distributions or rural works programs in the immediately affected areas (Tesfaye and others 1991, Webb and others 1992). Thin markets and lack of integration often necessitate separate parallel distribution systems for relief transport, storage, and distribution. These measures are likely to be both more effective in providing relief and more cost-effective than indirect fiscal and monetary measures.

The appropriate response of a donor must also depend strongly on the governance situation. If there is an effective government in which the donor has confidence, general support with financial or program food aid may still be an appropriate response to complement direct relief. Where a government is less effective or even nontherapeutic, the use of indirect channels is more appropriate. Examples appear to include Malawi and Mozambique in 1992–93, some of the Sahelian economies during the droughts of the early 1970s and 1980s, and the Horn of Africa up to the early 1990s.

Dualistic, Extractive Economies

An important issue is the extent to which those economies with relatively high GDP per capita need emergency assistance on any substantial scale. The governance issue also arises. Botswana,

Namibia, and South Africa would appear no longer to require substantial external assistance in response to a drought crisis. However, the interacting effects of conflict and economic decline leave the people of some countries with large extractive sectors vulnerable to any exogenous shock. Government and indigenous NGOs have to determine in consultation with their partners, the United Nations, and international NGOs whether the international community can play a useful complementary role in a drought crisis.

Additional Resources for Drought Crisis Management

The final issue to consider is whether the collective effort involving all these instruments and donors is commensurate with the scale of the financial problem created by a crisis. Unless there is an adequate, additional response to a drought in these terms, a heavy burden is imposed on both the public and private sectors of the affected economy, crowding out other expenditure and so implying developmental costs.

A good test case is the "successful response" to the drought in southern Africa in 1992 and 1993. The overall costs of relief operations in the region amounted to at least US$4 billion (Mugwara 1994). The foreign exchange costs of food imports and logistics were around US$1.6 billion, excluding South Africa. In addition, countries suffered loss of export earnings and governments suffered loss of revenue from the drought-related recession. However, the increase in total net aid disbursements during 1992 and 1993 from 1990 and 1991, again excluding South Africa, was slightly less than US$1.5 billion. Substantial increases in aid were already scheduled for Zambia and Zimbabwe to support structural adjustment. This suggests that the additional international resources, including all the food aid and emergency relief mobilized in response to the crisis, accounted for at most one-third of the direct costs of the drought relief program, without making any compensating allowance for the loss of export earnings. In addition, at least a third of the international response involved credits, thus, increasing the official debt.

The greater part of the drought response in southern Africa had to be financed by the affected countries, which reallocated expenditure within budgets already constrained by adjustment programs. Overall, a substantially greater international response on a grant basis would have contributed to reducing the negative developmental effects of the drought.

Drought Mitigation and Preparedness Strategies

Drought mitigation strategies have traditionally been defined in terms of improving food security at the national and household levels. Furthermore, such strategies have often been concerned with increasing long-run productivity rather than reducing output variability and so have only indirectly addressed problems of drought. Governments and donors have typically been most concerned about the possible implications of a further drought in the immediate aftermath of the last one. During this period, certain additional mitigation measures are discussed and sometimes implemented, including some investment in water resources. But, unless there is another drought, there has been little sustained interest by either the public or private sectors in mitigating the potential economic impacts of future disasters.

Farmers, particularly subsistence producers, have probably been the most consistently "drought-conscious" economic group in SSA. However, there is some evidence, for example, in southern Africa and even among subsistence farmers, that more widespread monocropping to increase productivity has gradually replaced traditional risk-averse, low-input, mixed farming systems over the past thirty years (World Bank 1995c). This trend, which is partly a consequence of deliberate policies, such as pan-territorial pricing for maize to enhance food self-sufficiency, has potentially increased vulnerability to drought.

There are, in fact, a number of ways in which preparedness measures can reduce the impacts of drought. This study has focused particularly on government and donor responses following the onset of a drought. Nevertheless, it also suggests conclusions in terms of appropriate and higher priority drought mitigation measures. These are detailed below, after reviewing a vital issue—the funding of such measures.

Governments in drought-prone countries can support drought mitigation measures both through direct investments and via policy instruments to influence the actions of the private sector, including industry and agriculture. The former entail clear opportunity costs in countries already facing difficult decisions in the allocation of scarce public resources. However, particularly in intermediate economies where governments may find themselves meeting increasingly large portions of any relief costs, the returns to drought investments may be high. Drought mitigation investments also entail capital outlays for the private sector, funded either by borrowing or plowing back profits. However, such expenditures are not necessarily high but may simply involve alternative choices of technology or modifications in existing practices.

The most important instruments available at the hands of the international community in supporting mitigation measures in drought-vulnerable countries include project and technical cooperation funds, SAP and other economic frameworks, and more general research-oriented funding. Donors may also be able to incorporate drought mitigation measures into development projects. In identifying and appraising potential projects, donors could also undertake assessments of drought vulnerability as part of the wider risk assessment proposal to help ensure that any necessary drought mitigation measures are built into projects and that project targets are not unrealistically optimistic. In terms of conditions attached to external assistance packages, any conditions regarding government spending need to be reviewed carefully to ensure that drought mitigation investments are not being seriously curtailed.

Understanding Drought Processes

There is a need to encourage links among technical, scientific, and social science research. The importance of this cannot be overstated (SADC 1993b, World Bank 1995c, Wilhite 1993)

Early Warning Systems

Early warning systems (EWSs) have developed using both narrowly technical and more general socioeconomic data to measure the vulnerability of African economies and societies and changing conditions. Taking as an example the 1991–92 drought in southern Africa, where technical systems were relatively good, a number of issues emerge relating to the nature and particular merits and limitations of such systems. EWSs gradually revealed the full scale of the drought impact over a period of four to five months, between November 1991 and April 1992, rather than immediately and so were most valuable when used in conjunction with flexible drought relief and rehabilitation programs.

EWSs are particularly important in providing information on the social dimension of droughts and are, therefore, especially important in ensuring policy sensitivity to smaller shocks and also for successful targeting. The organization of EWSs depends a lot on issues of governance, as well as differences of economic structure and resource endowments. The situation in the Sahelian countries is, for example, different from that in the Horn of Africa or the SADC region (Buchanan-Smith and Davies 1995).

Meteorological Forecasting and Climatic Change

The rapid and impressive advances in climatic research and weather forecasting systems are likely to provide earlier and more reliable predictions of drought events. When these developments are combined with the possibilities of computerized geographic information systems (GIS), there is considerable scope for pinpointing more precisely the areas of greater vulnerability and the drought risks impinging on economic activities and social groups.

A critical element in EWS is improved meteorological forecasts. These would effectively reduce some of the "shock" element of drought, assuming that they are updated frequently and that drought pronouncements are not made out of the blue. At least in the southern African context, considerable data already exist that are not being fully exploited (Gibberd and others 1995), suggesting scope for providing useful forecasts. However, climatologists also point to the deterioration in basic data from terrestrial stations in many parts of Africa, weakening the underlying informational basis of research and operational forecasting (Hulme 1995).

Furthermore, weather forecasts alone are not sufficient; information has to be made available in a form that is meaningful to key institutions and policymakers, including those responsible for public finances and monetary policies as well as those working in agriculture and water supplies. Moreover, these individuals require the analytical tools and the authority to exploit and respond appropriately and promptly to this information, whereas individual farmers require access to appropriate seed and other agricultural inputs. This point is critical. Information provided by EWSs has not always been heeded promptly, and there is no reason to believe that improved meteorological forecasts will be better used.[41] Further investigations of the reliability, use, and advantages of existing forecasts in the Sahel and equatorial East Africa could throw up some useful lessons on these issues.

Improved seasonal forecasts offer more tangible benefits, allowing farmers, at least in principle, to minimize losses by planting more drought-resistant crops or early maturing crops, depending on levels and timing of the rains (Gibberd and others 1995). However, these forecasts raise difficult dilemmas for governments in selecting appropriate economic policies. For example, precautionary stock building has real costs, and the imposition of an export ban can also upset prior commitments, threatening a country's reputation as a reliable supplier. Droughts can simultaneously deeply suppress effective demand, increase demand for credit, and exert heavy inflationary pressures on the economy, as demonstrated in the case of Zimbabwe in 1992–93. These impacts imply conflicting policy measures, the first two urgently require lower interest rates to help ensure early

---

[41] For example, despite meteorological information, including satellite data, as well as collation of physical data by several early warning networks following inadequate long rains (April–May) in Kenya in 1992, an international appeal was not made until October. Although low long rains in the following year indicated that the food situation would worsen, it was not until the end of 1994 that a further international appeal was made. Figures on the shortfall were not finalized until early 1995 (World Bank 1995a).

recovery of the economy, whereas the latter requires higher rates. In practice, governments have often continued to pursue existing policies, but the benefits of adjusting policies is an area that needs further research.

The distinctive and systematic country and regional rainfall patterns revealed by climatic research imply not only different probabilities of extreme drought events but that these are subject to cyclical and secular changes. The recent secular aridification in Saharan countries is well recognized. There is some potential for further aridification, for example, in southern Africa; modeling of climatic change indicates that extreme drought events of the intensity experienced in the past two decades may become more frequent (Hulme 1996). Such evidence could be used to formulate better rules of thumb for both macroeconomic sectoral and provincial level planning.

The advent of computerized GISs facilitates the use of combined climatic, soil, and topographical data. These can also be integrated with data on economic activities and potentially vulnerable social groups. There is considerable opportunity, therefore, for drought risk mapping that could provide a basis for preparedness, including drought mitigation investments, as well as assisting the better targeting of crisis response measures.

Economic Planning

The possibility of drought has typically been ignored in medium-term planning. This probably implies that overoptimistic growth targets have been set, drought contingency plans have not been made, and the risk of drought has not been adequately incorporated into budgetary exercises, sectoral strategies, or investment programming. The risk of drought should be explicitly acknowledged and built into overall development strategies and sectoral plans as well as into SAPs. Drought sensitivity analysis should also be conducted before promoting particular sectors or subsectors, particularly as part of any diversification strategy. Furthermore, sensitivity analysis could systematically explore the implications of climatic change for investment and wider resource allocation, in particular, increases in the frequency of events beyond the range of currently normal risk management practice. These suggestions are reflected in the recommendations on areas for further research below in chapter 7.

Water Resource Management

For governments confronting the intense financial pressures of adjustment and growing demands for water the challenge is to maintain a substantial cushion of noncritical use under conditions of normal precipitation (Frederiksen 1992). This issue cannot be disassociated from that of investment in drought proofing. Water tariffs and hydroelectric power supplies should reflect the long-run marginal cost of supply, taking account of new, possibly increasingly costly investments to meet future rises in demand. This would encourage users to avoid excessive use of water, a practice that will be increasingly important as expanding populations and increasing economic activity create additional demands. Further investments should also be made in water resource development, whereas reductions in waste and physical loss, recycling of water, and improvements in sanitation should be promoted (Winpenny 1994). The hydrological situation also needs to be monitored carefully and continuously and appropriate measures taken in light of evidence on changes resulting from the direct effects of human activity interacting with climatic trends.

Drought Insurance

In SSA drought insurance is limited, both within agriculture and in water-intensive industries more generally.[42] Insurance is not an economic solution to potential disasters but simply a means by which to transfer risk, changing the nature of the financial costs and so the economic impacts of drought but not necessarily reducing them. Nevertheless, more widespread insurance could encourage the adoption of drought mitigation or water conservation strategies within the industrial and commercial sectors to the extent that policies were conditional on certain measures being taken or that premiums reflected levels of exposure to drought. The response of farmers to drought insurance is less clear. For example, in Australia provision of drought relief has effectively meant that farmers have had little incentive to protect themselves against future droughts (Allan and Heathcote 1987). Large-scale commercial farmers in Namibia and South Africa also appear to have adopted farming systems that do not fully internalize drought risk (Namibia 1997, Rimmer 1996). However, this does not rule out benefits of public assistance, including assisting a speedy recovery.

The way forward may be to encourage commercial agricultural and water-intensive industries more generally to internalize risks in both their production decisions and choice of technology through investment. This might involve both ending nearly automatic drought relief and supporting the establishment of self-financing insurance mechanisms in countries such as Namibia and South Africa. From a commercial insurer's point of view, the fact that severe droughts would result in heavy claims does not necessarily preclude the provision of such coverage. Admittedly, reinsurance is becoming more and more costly for certain groups, because rates are increasingly calculated on the basis of actual risk in specific circumstances rather than on global or regional averages. However, there are creative possibilities that could be explored. For example, it has been suggested on several occasions that disaster insurance exchange schemes should be set up among countries in different climatic or geological regions.[43] Such schemes and the implications for the economywide impacts of drought in spreading and altering the nature of risk are an important area for further research.

Agricultural and Food Policies

This study has focused deliberately on the nonagricultural aspects of drought. In part, this was because of a perceived imbalance in previous work. There are, however, context-specific issues of importance regarding agriculture including:

Minimum Essential Food Stocks at a National and Decentralized Level

On the basis of the currently limited availability of reliable meteorological forecasts, drought-prone countries should continue to maintain some level of cereals reserves, as demonstrated by the experience of Zambia and Zimbabwe in 1992. However, individual countries need to review the acceptable level of risk and appropriate size of reserves. These are costly to maintain, and countries may, therefore, consider holding part of the reserve in financial form instead. This funding, held in

---

[42] For example, there is some form of crop insurance available in Zambia. In 1992 the Zambian State Insurance Company estimated that claims from more than 9,500 farmers would exceed Ksh600 million (Mulwanda 1995). The record of falling crop insurance schemes in developing countries has been almost uniformly negative (Wright and Hewitt 1993).

[43] For example, such a scheme has been suggested by the Alliance of Small Island States, although it has not actually been implemented (IIED 1992).

domestic currency or foreign exchange or both, depending on specific country circumstances, could be used to purchase food imports as required or, as suggested by the World Bank (1995c), placed in a revolving fund to support private sector import activity. Governments could also elect to set aside contingency funds for more general drought emergency purposes.

Market Liberalization

Across much of SSA, producer and consumer prices are being decontrolled and agricultural markets deregulated. This potentially permits earlier market responses to drought, dampening large price increases and reducing the need for large-scale provision of food aid. However, increasing liberalization of food imports across SSA could imply that El Niño/Southern Oscillation–related droughts result in increasing demand for commercial imports. Such policies also need to be complemented by carefully targeted interventions to support the most vulnerable groups who will not be able to procure food at (probably) inflated prices during periods of drought. In addition, careful coordination is also required to ensure that the government and donors are able to monitor scheduled food imports and any potentially food deficit not covered during a drought. Evidence suggests that governments should resist the temptation to reimpose controls, such as an export ban or consumer price subsidies, in a drought emergency.

Diversification of Crops

It is important to ensure that crops are suitable to local growing conditions. In some countries this may entail a reversal of policies that have promoted the production of food staples such as maize at the expense of more stress-tolerant millets, which may be more appropriate in areas of more marginal rainfall. A reversal of gradual trends toward monocropping back to intercropping of more drought-tolerant crops could also reduce vulnerability to drought.

Crop Varietal Development and Promotion

There is potential to improve the productive potential of more stress-tolerant varieties and make these plant types widely available.

Irrigation

Most of the irrigation in SSA depends on interannual storage and suffers from poor system management, rendering it highly drought sensitive. There are, therefore, difficult issues that need to be confronted in balancing returns to further investment and better management of available supplies. Furthermore, future growth in agricultural productivity in SSA will depend partly on the expansion of irrigation.

Overall, there is abundant evidence that traditional farming and pastoral systems were well adapted to rainfall variability, apart from infrequent extreme events. The direction of change under demographic pressure and the forms of intensification of production encouraged by public policy may have increased the exposure of many systems. The challenge in a more liberal economic environment is to find ways of both enabling agricultural production to expand, while internalizing risks and buffering the most vulnerable against unacceptably high social impacts.

Drought Contingency Planning

The droughts of the early 1970s and early 1980s gave impetus to the CILSS[44] and Club du Sahel initiatives. The mid-1980s crisis resulted in the Inter-Governmental Authority on Drought and Development (IGADD) involving countries of the "Greater Horn" of Africa. Since the 1991–92 drought, governments, the IFIs, and bilateral donors have accorded an increased emphasis to contingency planning for southern Africa (e.g., SADC 1993a, Callihan and others 1994, World Bank 1995c). Discussion following that emergency highlights some experiences relevant to future contingency plans:

- The appropriate role of food security stocks or financial contingency reserves is unresolved partly because market liberalization and greater regional integration are changing the parameters for future contingency planning (SADC 1993b).

- Postdisaster recovery, including seed supply and livestock rehabilitation, is generally poorly handled (ODI 1997, Friis-Hansen and Rohrbach 1993).

- Emergency, water supply interventions are widely ineffective in assuring or augmenting supplies during a drought and are not properly integrated with health measures (Taylor and Mudege 1992, Clay and others 1995, Mason and Leblanc 1993).

- Credit mechanisms are often not in place to assist affected producers.

- Employment-generating, livelihood-sustaining rural works activities have an important potential role in highly drought-vulnerable areas and also as part of antirecessionary measures (Webb and others 1992, Webb and Moyo 1992).

- Drought mitigation needs to be incorporated into long-term poverty programs, including rural employment and income and food transfer activities. But this should be done in ways that do not create unsustainable welfare dependence in more vulnerable areas and groups. Furthermore, the prevalent assumption that these should be food for work or food-based income-supplementing programs should be challenged. In increasingly better integrated and more complex systems, cash for work, food coupons, and other cash transfer mechanisms may be more efficient than food-based interventions (Drèze and Sen 1989).

Overall, there are many things that can be considered in terms of drought mitigation at different levels:

---

[44] Comité permanent inter états de lutte contre la sécheresse au Sahel (Interstate Committee for the Fight against Drought in the Sahel).

- National, highly context-specific activities

- Regional issues, including meteorological forecasting, EWSs and logistics systems

- Some Africawide issues, including in particular further technical and multidisciplinary work on the understanding of drought processes and their policy implications.

CHAPTER 7:

CONCLUSIONS

The following are the main conclusions and policy implications arising from this paper.

Economywide Impacts of Drought

- Drought shocks have large, but highly differentiated, economywide impacts. The likely frequency, scale, and character of these impacts depend on the interaction of economic structure and resource endowments, as well as on more immediate short-term economic factors.

- Counterintuitively, some of the relatively more developed or "complex" SSA economies, such as Senegal, Zambia, and Zimbabwe, may be more vulnerable economically to drought shocks than less developed and more arid countries, such as Burkina Faso, or than countries such as Somalia that are undergoing conflict-related emergencies. As economies become more complex and diversified, they become less vulnerable to drought. This suggests an "inverted U"–shaped relationship between the level of complexity of an economy and its vulnerability to drought.

- Regardless of differing economywide consequences, droughts in SSA invariably have severe negative food security implications for certain segments of the population. These impacts occur at various levels in economies (household, sectoral and regional).

- In the least developed countries, drought should be viewed and treated less as an exogenous shock, as is currently done, and more as a structural problem.

Climatological Issues

- This study emphasizes the importance of resource endowments, to drought vulnerability, but does not examine climate change issues in any detail. However, the actual rainfall patterns in the Sahel and southeast Africa in the last twenty years suggest some increases in drought risk; the modeling of long-run global climatic scenarios provides no basis for complacency on this issue.

- Further climatic research is essential to strengthening understanding of the forcing mechanisms behind rainfall variability. This research will require modeling as well as statistical analysis. At the level of individual countries and agroecological regions, further investigations of the interseasonal variability of rainfall and identification of patterns and possible changing relations would contribute to better understanding of the agroclimatic aspects of drought and provide a building block for more refined meteorological forecasting.

- An investigation of the reliability, extent of use, and benefits of existing meteorological forecasts provided in the Sahel and equatorial East Africa would help clarify the merits of investing substantial resources in the development of improved weather forecasts elsewhere in SSA. It could also offer useful lessons on the types of forecast information that could be most usefully provided.

Drought Mitigation and Response

- Both strategies to mitigate the impacts of drought in the long term and effective responses to specific shocks have to be sensitive to differences in economic structure, resource endowment, and prevailing economic circumstances. Policies are likely to be poorly calibrated if they are based on Africawide or even more general prescriptions for drought mitigation.

- The articulation of strategies for distinct subregions finds support in the current state of scientific understanding of climatic variability. The long-run problem of aridification in the Sahel, the quasi-cyclical weather patterns of southeast Africa, and the classically random patterns of East Africa are broadly coterminous with the CILSS, SADC and IGADD regions. However, from a climatic viewpoint, at least two large countries, Sudan and Tanzania, should be more involved in regional drought mitigation policy with CILSS and IGADD countries respectively. This underlines the need for understanding country-specific patterns and developing predictors that can be used in economic management and long-term planning.

- In responding to a drought, financial aid for balance of payments and budgetary support should have the highest priority in more complex SSA economies, although carefully targeted food aid interventions may also be required. Large-scale targeted interventions should be the primary modality of response in simpler and conflict-affected economies. Targeted interventions may also be required in dualistic economies with large extractive sectors. The appropriate balance of response in all three cases also depends on the quality of governance in the drought-afflicted economies. Taking a long-term perspective, the processes of increasing integration of economies more widely and liberalizing food markets imply a shift of emphasis to financial assistance more generally and the use of cash-based targeted interventions that sustain livelihoods rather than supporting food security with relief food.

- There has typically been little sustained interest in drought mitigation on the part of either governments or donors, except in terms of improving food security. Any such strategies have traditionally been defined in terms of improving food security at the national and household levels, often focusing on the production of only one food crop. However, there are a number of other ways of reducing the impact of a drought through, for example, improved water conservation and management, increased planting of drought-tolerant plants, and ensuring that the risk of drought has been adequately built into strategies for promoting economic diversification.

- Economic investment and water resource management strategies should be formulated on the basis of the best available *long-term* scientific data for a particular region. This conclusion is illustrated by the increasing vulnerability of the Kariba Zambezi system to extreme climatic conditions from the mid-1980s.

- Economic drought mitigation, relief, and rehabilitation instruments should not be regarded as separate, more or less autonomous areas of policy action. Instead, donors both collectively and individually should seek to ensure the better integration and greater overall coherence of individual programs and policies. The most appropriate and also the most likely balance of actions (which are not necessarily the same) will depend on the country policy environment.

- There are likely to be two overlapping monitoring systems concerned with food security and economic issues. The former would be more effective if information on rainfall and the agricultural and food situation were more regularly integrated into economic monitoring. Similarly, those monitoring economic performance should take a closer interest in the interpretation and use of technical and scientific information on rainfall.

Drought and Structural Adjustment Programs

- A weakness in recent policy practice has been a failure to take potential drought shocks into account in formulating medium-term economic strategies, for example, in the formulation of SAPs. This study gives further support to the argument that the formulation of SAPs should be sensitive to drought. It is not clear that this implies redefining medium-term policy goals, assuming they are based on a realistic assessment of the economic environment rather than the most favorable assumptions. Rather, it implies that there ought to be some element of preparedness. In designing SAPs, thought should be given to what would be required if there were to be a severe drought or, indeed, some other major economic shock. This exercise might be based in part on exploring, as has been done in this study, one or more recent shocks. On a positive note, since the 1991–92 drought, there are indications in southern Africa of serious attempts to rectify this omission.

- The cases examined suggest that when a drought shock occurs in the context of a SAP some combination of the following responses may be appropriate:

- Rephasing of some policy objectives

- Redirection of already committed resources, underutilized because of the shock

- Rapid commitment and *disbursement* of additional external resources.

- Sustaining a SAP should not oblige a government to finance its drought response in ways that intensify the recessionary effects of the shock on the domestic economy.

Areas for Further Economic Research

- The impacts of drought mediated through intersectoral linkages are complex. Without more sophisticated modeling, these effects could be addressed only superficially in this study. However, the study has suggested a broad framework within which some of these issues can be explored in more detail at a country-specific level.

- The complexities of economic structure and resource endowment justify closer exploration of the dynamics of highly drought-vulnerable economies through economic models. The objective of modeling would be to inform broader policy with a better understanding of the consequences of economic changes rather than to fine-tune short-term economic policy. It could be most beneficial to explore ways in which existing models employed in economic forecasting and policy analysis—for example, the World Bank Revised Minimum Standard Models (RMSMs)—might be adapted to take account of drought. Further research is also required on the underlying behavioral relationships between drought shocks and economic and policy variables, both as a prerequisite for any modeling of drought shocks and to contribute toward a better understanding of appropriate policy responses.

- The effects of drought on public expenditure should be explored in detail through country studies. For example, how have governments financed additional expenditures or handled reallocation of expenditures and with what consequences?

- Drought shocks clearly have important implications for the pace of implementation and success of SAPs. This study has considered some preliminary evidence of the intricate and complex relationship between the two and suggested some scope for offsetting any adverse impacts. However, further research is needed to provide governments and policymakers with careful advice about how to respond to the risk and actual occurrence of drought shocks during periods of economic reform.

- The modeling of drought impacts should take into account established regional rainfall patterns, probably by investigating countries from each of the three regions distinguished by climatological research—the Sahel, southeast Africa and East Africa, and the Horn. There are also three complementary sets of scenarios that would be useful to explore. First, economic performance could be projected on the basis of parametric values reflecting "normal" or mean rainfall. Second, rainfall variability could be built into the simulation analysis as a random process but reflecting the specific rainfall pattern of each region. Subsequently, it will be especially useful to discover whether the results of this more complex modeling could be closely approximated by some more simple rule, such as, for example, one serious drought event every five years. Third, the efficacy of meteorological forecasting should be investigated. This might, for example, be done by taking forecasts for a specific region and examining the consequences of correct and incorrect forecasts over the range of wet to dry quintiles currently produced. This would require the introduction of some simple decision rules on, for example, cereal stocks and food imports into the modeling.

Ultimately, there is no single set of policies or mitigation and relief measures that can be implemented anywhere and at any time to combat the impacts of drought. Success stories have owed much to chance occurrences, such as high commodity export prices, to specific country circumstances, as well as to deliberate policies and actions. Nevertheless, there are lessons to be learned from country experiences in developing strategies to reduce the economywide impacts of droughts.

REFERENCES

"Adjustment in Africa: Reforms, Results, and the Road Ahead—The Zambian Experience." 1994. Paper presented at the World Bank Seminar on Adjustment in Africa, Harare.

Ainsworth, M. and M. Over. 1992. "The Economic Impact of AIDS: Shocks, Responses, and Outcomes." In M. Essex and others, eds. *AIDS in Africa*. New York: Raven Press.

———. 1994. "AIDS and African Development." *World Bank Research Observer* 9(2): 203–40.

Allan, R and R. L. Heathcote. 1987. "The 1982–3 Drought in Australia." In M. Glantz, R. Katz, and M. Krenz, eds. *Climate Crisis: the Societal Impacts Associated with the 1982–83 Worldwide Climate Anomalies*. Boulder (Colorado) and Nairobi: National Center for Atmospheric Research and United Nations Environment Program.

Banda, A. K. 1993. "Country Assessment on the Drought Situation in Zambia." Paper presented at SADC Regional Drought Management Workshop, Harare, September 13–16. Ministry of Agriculture, Food, and Fisheries, Policy and Planning Division, Lusaka.

Benson, C. 1994. "Drought and Macroeconomic Performance: A Comparative Analysis of Six Sub-Saharan African Economies." Mimeo. Overseas Development Institute, London.

———. 1997. "Drought and the Zimbabwe Economy 1980–93." In H. O'Neill and J. Toye, eds., forthcoming, *A World Without Famine?* Basingstoke: Macmillan.

Benson, C. and E. J. Clay. 1994a. "The Impact of Drought on Sub-Saharan African Economies." *IDS Bulletin* 25(4):24–32.

———. 1994b. "The Impact of Drought on Sub-Saharan Economies: A Preliminary Examination." ODI Working Paper 77. Overseas Development Institute, London.

Berg, E. 1975. "The Economic Impact of Drought and Inflation in the Sahel." Based on *The Recent Economic Evolution of the Sahel*. Prepared by the University of Michigan for the U.S. Agency for International Development. University of Michigan, Ann Arbor.

Block, S. and C. P. Timmer. 1994. "Agriculture and Economic Growth: Conceptual Issues and the Kenyan Experience." Harvard Institute for International Development, Cambridge, MA.

Bonitatibus, E. and J. Cook. 1996. *Incorporating Gender in Food Security Policies: A Handbook for Policymakers in Commonwealth Africa*. London: Commonwealth Secretariat.

Borton J. and others 1988. "ODA Emergency Aid to Africa 1983–86." Evaluation Report EV425. Overseas Development Administration, London.

Buchanan-Smith, M. and S. Davies. 1995. *Famine Early Warning and Response: the Missing Link*. London: Intermediate Technology.

California Department of Water Resources. 1991. *California's Continuing Drought 1987–91*. Sacramento, Calif.

Callihan, D. M. and others. 1994. "Famine Averted: the United States Government Response to the 1991–92 Southern African Drought." Management Systems International, Washington, D.C.

Chen, M. 1991. *Coping with Seasonality and Drought*. New Delhi: Sage Publications.

Clay, E. J. 1994. "Drought in Southern Africa: Shocks and Research." *Development Research Insights* 12:1–2. Overseas Development Institute, London and Institute of Development Studies, Brighton.

———. 1997. "Responding to the Human and Economic Consequences of Natural Disasters." In H. O'Neil and J. Toye, eds. *A World Without Famine?* Basingstoke: Macmillan.

Clay, E. J. and others 1995. "Evaluation of ODA's Response to the Southern Africa Drought." Evaluation Report EV568. Overseas Development Administration, Evaluation Department, London.

Clay, E. J., S. Dhiri, and C. Benson. 1996. *Joint Evaluation of European Union Programme Food Aid*. London: Overseas Development Institute.

Collier, P. and J. W. Gunning. 1996. "Policy Toward Commodity Shocks in Developing Countries." *IMF Working Paper* No WP/96/84. International Monetary Fund, Washington, D.C.

Collins, C. 1993. "Famine Defeated: Southern Africa, U.N. Win Battle Against Drought." *Africa Recovery Briefing Paper* 9. United Nations Department of Public Information, New York.

Cuddington, J. T. 1991. "Modeling the Macroeconomic Effects of AIDS with an Application to Tanzania." Department of Economics Working Paper No. 91-17. Georgetown University, Washington, D.C.

Cuddington, J. T. and J. Hancock. 1992. "Assessing the Impact of AIDS on the Growth Path of the Malawian Economy." Department of Economics Working Paper No 92-07. Georgetown University, Washington, D.C.

Davies, R. 1992. "Macroeconomic Aspects of Zimbabwe's 'Transition from Socialism.'" University of Zimbabwe, Department of Economics, Harare.

Davies, R., J. Rattso, and T. Ragnar. 1993. "The Macroeconomics of Zimbabwe in the Eighties—A CGE-Model Approach." University of Zimbabwe, Department of Economics, Harare.

Davies, S. 1995. *Adaptable Livelihoods: Coping with Food Insecurity in the Malian Sahel*. Basingstoke: Macmillan.

Delgado, C. and others. 1994. "Agricultural Growth Linkages in Sub-Saharan Africa." U.S. Agency for International Development, Washington D.C.

Devereux, S. and others. 1995. *The 1992–93 Drought in Namibia: An Evaluation of its Socio-Economic Impact on Affected Householders*. Windhoek: Gamsberg Macmillan.

DHA and SADC. 1992. "Drought Emergency in Southern Africa (DESA): Consolidated UN-SADC Appeal Mid-Term Review." U.N. Department of Humanitarian Affairs, Geneva.

Downing, T. E., K. W. Gitu, and C. M. Kamau. 1987. *Coping with Drought in Kenya: National and Local Strategies.* Boulder and London: Lynn Rienner Publishers.

Drèze, J. and A. Sen. 1989. *The Political Economy of Hunger.* Oxford: Clarendon Press.

Elbadawi, I. A. 1997. "Structural Adjustment and Drought in Sub-Saharan Africa." In H. O'Neill and J. Toye, eds. *A World Without Famine?* Basingstoke: Macmillan.

Elbadawi, I., D. Ghura, and G. Uwujaren. 1992. "Why Structural Adjustment Has Not Succeeded in Sub-Saharan Africa." Policy Research Working Paper 1000. World Bank, Washington D.C.

Food and Agriculture Organization (FAO). 1978. "Report on Agro-Ecological Zones Project." *Methodology and Results for Africa. World Soil Resources Report* 48 (1). Rome.

————. 1996. *Rome Declaration on World Food Security and World Food Summit Plan of Action.* World Food Summit, November 13–17, 1996. Rome.

Frederiksen, A. D. 1992. Drought Planning and Water Efficiency Implications in Water Resource Management." World Bank Technical Paper Number 185. World Bank, Washington D.C.

Friis-Hansen, E. and D. D. Rohrbach. 1993. "SADC/ICRISAT 1992 Drought Relief Emergency Production of Sorghum and Pearl Millet Seed: Impact Assessment." Working Paper 93/01. International Crops Research Institute for the Semi-Arid Tropics, Southern and Eastern Africa Region, Bulawayo.

Gibberd, V. and others. 1995. "Drought Risk Management in Southern Africa: The Potential of Long Lead Climate Forecasts for Improved Drought Risk Management." Natural Resources Institute, Chatham.

Glantz, M. H., ed. 1987. *Drought and Hunger in Africa: Denying Famine a Future.* Cambridge: Cambridge University Press.

Hannover, W. and others 1996. "Kenya: A Rapid Evaluation." Joint Evaluation of EU Programme Food Aid. ODI, London.

Hicks, D. 1993. "An Evaluation of the Zimbabwe Drought Relief Programme 1992–1993: The Roles of Household Level Response and Decentralized Decision Making." World Food Programme, Harare.

Hogan, L. and others 1995. "The Impact of the 1994–95 Drought on the Australian Economy." Paper presented at the ABARE Conference on Coping with Drought, Australian Institute of Agricultural Science 1995 Drought Forum, New South Wales, April 6–7. Australian Bureau of Agricultural and Resource Economics, Canberra.

Hulme, M. 1992. "Rainfall Changes in Africa: 1931–60 to 1961–90." *International Journal of Climatology* 12: 685–99.

————. 1995. "Climatic Trends and Drought Risk Analysis in Sub-Saharan Africa." University of East Anglia, Climatic Research Unit, Norwich.

————. ed. 1996. "Climate Change and Southern Africa: An Exploration of Some Potential Impacts and Implications for the SADC Region." Report commissioned by WWF International. Climatic Research Unit, University of East Anglia, Norwich.

International Fund for Agricultural Development (IFAD). 1994. "Development and the Vulnerability of Rural Households to Drought: Issues and Lessons from Sub-Saharan Africa." Paper for Technical Session: Managing Drought, World Conference on Natural Disaster Reduction, Yokohama, May 23–27, 1994. Africa Division, Rome.

International Institute for Environment and Development (IIED). 1992. "International Insurance Pool Proposed." *Tiempo* 4. London.

Jean, F. 1993. "Somalia: Humanitarian Aid Outgunned." In F. Jean, ed. *Life, Death, and Aid*. London: Routledge.

Jones, E. L. 1988. "Natural Disasters and the Historical Response." In *Australian Economic History Review* 28 (1) (March). Melbourne: Oxford University Press Australia.

Kebbede, G. 1992. *The State and Development in Ethiopia*. New Jersey/London: Humanities Press.

Killick, T. and M. Malik. 1992. "Country Experiences with IMF Programmes in the 1980s." *The World Economy* 15 (5): 599–632.

Kox, H. 1990. "Export Constraints for Sub-Saharan Growth: the Role of Non-Fuel Primary Commodities." Research Memorandum 1990–91. Faculty of Economics and Econometrics, Free University. Amsterdam.

Krugman, P. 1988. "External Shocks and Domestic Policy Responses." In R. Dornbusch and F. L. C. H. Helmers, eds. *The Open Economy: Tools for Policymakers in Developing Countries*. Oxford: Oxford University Press.

Kuznets, S. 1955. "Economic Growth and Income Inequality." *American Economic Review* 45: 1–28.

Le Houérou, H. N. and others 1993. "Agro-Bioclimatic Classification of Africa." Agrometeorology Series Working Paper 6. Food and Agriculture Organization, Rome.

Legal, P. Y. and others 1996. "Zambia: A Rapid Evaluation." Joint Evaluation of EU Programme Food Aid. ODI, London.

Lindesay, J. A. and C. H. Vogel. 1990. "Historical Evidence for Southern Oscillation-Southern African Rainfall Relationships." *International Journal of Climatology* 10: 679–90.

Macrae, J. and A. Zwi. 1994. War and Hunger: Rethinking International Responses to Complex Emergencies. London: Zed Books.

Mason, J. P. and M. Leblanc. 1993. "Evaluation of Africare Emergency Water Relief Regional Project: Zimbabwe, Malawi, Zambia." U.S. Agency for International Development, Office of Foreign Disaster Assistance, Disaster Response Division, Washington D.C.

Mason, S. J. and P. D. Tyson. 1992. "The Modulation of Sea Surface Temperature and Rainfall Associations over Southern Africa with Solar Activity and the Quasi-Biennial Oscillation." *Journal of Geophysical Research* 97: 5847–56.

Mulwanda, M. 1995. "Structural Adjustment and Drought in Zambia." *Disasters* 19(2): 85–93 .

Mugwara, R. 1994. "Linking Relief with Development in Southern Africa: an SADC Perspective on the 1991–92 Drought Emergency." *IDS Bulletin.* 24(4): 92–6.

Namibia. 1997. "Toward a Drought Policy for Namibia." Papers prepared for National Workshop on Drought Policy. Windhoek.

Nowlan, J. and B. Jackson. 1992. "Drought in Southern Africa." Version 5. U.S. Agency for International Development, Harare.

Overseas Development Institute (ODI). 1997. "Seed Provision During and After Emergencies." *Good Practice Review* 4, Relief and Rehabilitation Network, London.

Pretorius, C. J. and C. J. Smal. 1992. "Notes on the Macroeconomic Effects of the Drought." *South African Reserve Bank Quarterly Bulletin* 184 (June): 31–8.

Puetz, D., S. Broca, and E. Payongayong, eds. 1995. "Making Food Aid Work for Long-Term Food Security: Future Directions and Strategies in the Greater Horn of Africa." Proceedings of a USAID/IFPRI Workshop, March 27–30, Addis Ababa.

Purtill, A. and others. 1983. "A Study of the Drought." Supplement to the *Quarterly Review of the Rural Economy* 5(1): 3-11.

Reardon, T. and others. 1988. "Coping with Household-Level Food Insecurity in Drought-Affected Areas of Burkina Faso." *World Development* 16(9): 1065–74.

Reardon, T. and J. E. Taylor. 1996. "Agroclimatic Shock, Income Inequality, and Poverty: Evidence from Burkina Faso." *World Development* 24(5): 901–14.

Relief and Rehabilitation Commission (RRC). 1985. "The Challenges of Drought: Ethiopia's Decade of Struggle in Relief and Rehabilitation." Addis Ababa.

Riley, B. 1993. "A Report on the Status of Food Insecurity in Zambia." Draft. World Bank, Washington, D.C.

Rimmer, M. 1996. "Managing Drought in South Africa, Issues from the 1995–96 Drought Relief Programme." Land and Agriculture Policy Center, Johannesburg.

Robinson, P. 1993. "Economic Effects of the 1992 Drought on the Manufacturing Sector in Zimbabwe." Overseas Development Institute, London.

Rowell, D. P. and others. 1995. "Variability of Summer Rainfall over Tropical North Africa (1906–92): Observations and Modeling." *Quarterly Journal of the Royal Meteorological Society* 121: 669–704.

SADC. 1993a. "Assessment of the Response to the 1991–92 Drought in the SADC Region." Southern African Development Community, Food Security Technical and Administrative Unit, Harare.

———. 1993b. "Drought Management Workshops in Southern Africa." Report of the SADC Regional Drought Management Workshop, Harare. September 13–16. Southern African Development Community, Food Security Technical and Administrative Unit, Harare.

Scoones, I. and others. 1996. *Hazards and Opportunities: Farming Livelihoods in Dryland Africa, Lessons from Zimbabwe*. London: Zed Press.

Seshamani, V. 1993. "The Role of the International Community in Drought Management in Zambia. From Recent Experience to Future Preparedness." Annex to Zambia Country Assessment paper presented at SADC Regional Drought Management Workshop, Harare, September 13–16. University of Zambia, Department of Economics, Lusaka.

Sheets, H. and R. Morris. 1974. *Disaster in the Desert: Failure of International Relief in West African Drought*. Washington, D.C.: Carnegie Endowment for International Peace.

Stewart, F. 1993. "War and Underdevelopment: Can Economic Analysis Help Reduce the Costs?" *Journal of International Development* 5(4): 357–80.

Taylor, P. and N. R. Mudege. 1992. "Emergency Drought Relief Programmes in Zimbabwe. Lessons for Today from Past Experience." Training Research Center for Water and Sanitation, University of Zimbabwe, Harare.

Tesfaye, T. and others. 1991. "Drought and Famine Relationships in Sudan: Policy Implications." *Research Report* 88. Washington DC: IFPRI.

Thompson, C. 1993. "Drought Emergency in Southern Africa: The Role of International Agencies." Paper presented at SADC Regional Drought Management Workshop, Harare, September 13–16. University of Zimbabwe, Harare.

Thomson, A. 1994. "The Impact of Drought on Government Expenditure in Namibia in 1992–93." Overseas Development Institute, London.

———. 1995. "Drought and Market Liberalization in Kenya." Overseas Development Institute, London.

Tiffen, M. and others. 1994. *More People, Less Erosion: Environmental Recovery in Kenya*. Chichester: John Wiley.

Tiffen, M. and M. R. Mulele. 1993. *The Environmental Impact of the 1991–2 Drought on Zambia*. Gland: International Union for the Conservation of Nature and Natural Resources.

Tobaiwa, C. 1993. "Zimbabwe: The Response to the 1992 Drought in the Context of Long-Term Development Objectives." Paper presented at SADC Regional Drought Management Workshop, September 13–16, Harare.

Tschirley, D., C. Donovan, and M. T. Weber. 1996. "Food Aid and Food Markets: Lessons from Mozambique." *Food Policy* 21(2): 189–210.

Tyson, P. D. and T. G. Dyer. 1978. "The Predicted Above-Normal Rainfall of the Seventies and the Likelihood of Droughts in the Eighties in South Africa." *South African Journal of Science* 74: 372–77.

United Nations Development Programme (UNDP). 1995. *Human Development Report 1995*. New York and Oxford: Oxford University Press.

Van der Hoeven, R. 1994. "External Dependence, Structural Adjustment and Development Aid in Sub-Saharan Africa." Paper presented at International Policy Workshop on International Capital Flows and Economic Adjustment, Institute of Social Studies, The Hague, December 2–3. International Labor Organization, Geneva.

Webb, P. and S. Moyo. 1992. "Food Security through Employment in Southern Africa: Labor-Intensive Programs in Zimbabwe." IFPRI, Washington D.C.

Webb, P. and others 1992. "Famine in Ethiopia: Policy Implications of Coping Failure at National and Household Levels." *Research Report* 92. Washington D.C.: IFPRI.

Wilhite, D. A. ed. 1993. *Drought Assessment, Management, and Planning: Theory and Case Studies*. Boston: Kluwer Academic Publishers.

Winpenny, J. 1994. *Managing Water as an Economic Resource*. London: Routledge.

World Bank. 1985a. "Project Performance Audit Report. Ethiopia Drought Areas Rehabilitation Project." (Credit 485-ET.) Report No. 6018. Operations Evaluation Department. Washington, D.C.

———. 1985b. "Report and Recommendation of the President of the International Development Association to the Executive Directors on a Proposed Credit in an Amount Equivalent to SDR 31.3 Million to Ethiopia for a Drought Recovery Programme." Washington, D.C.

———. 1988. "Adjustment Lending: An Evaluation of Ten Years of Experience." Washington D.C.

———. 1991a. "Ethiopia. Drought Recovery Program. (Credit 1576-ET)." Project Completion Report. Report No. 8639. Agriculture Operations Division, Country Department II, Africa Regional Office, Washington, D.C.

———. 1991b. "Food Security and Disasters in Africa: A Framework for Action." Africa Technical Department, Washington. D.C.

———. 1991c. "Food Security and Slow Onset Disasters in Eastern Africa: Departmental Action Plan." Eastern Africa Department, Agriculture Operations Department, Washington D.C.

———. 1993. *World Development Report 1993*. New York: Oxford University Press.

———. 1994. *Adjustment in Africa: Reforms, Results, and the Road Ahead*. World Bank Policy Research Paper. Oxford: Oxford University Press.

———. 1995a. "1995 Drought Impact and Implications for Southern Africa: A Preliminary Assessment." Paper presented at Inter-Agency Meeting on the 1995 Drought in Southern Africa, Paris, March 15. Southern Africa Department, Washington D.C.

————. 1995b. *Kenya Poverty Assessment*. Report No. 13152-KE. Eastern Africa Department, Population and Human Resources Division, Washington, D.C.

————. 1995c. "Southern Africa 1995: Drought Vulnerability, Drought Mitigation, and Long-Term Development Strategies." Southern Africa Department, Agriculture and Environment Division, Washington, D.C.

World Food Programme (WFP). 1993. "A Disaster Averted." *World Food Programme Journal* 25: 22–24.

————. 1994. "Terminal Evaluation: Summary Report on Umbrella Operation on Southern African Drought." Emergency Operation No. 5052-60, CFA/37/C SCP: 12/6-B1. Rome.

Wright, B. D. and J. A. Hewitt. 1993. "Crop Insurance for Developing Countries." In P. Berck and D. Bigman, eds. *Food Security and Food Inventories in Developing Countries*. Wallingford: CAB International.

Zimbabwe, Ministry of Public Service, Labor, and Social Welfare. 1995. "A State of Disaster: Government of Zimbabwe Appeal for Assistance." Harare.

APPENDIXES

Appendix Table 1: Results of OLS regression analysis on the relationship between GDP and agricultural sector growth and drought (a)

		GDP		Agriculture				Manufacturing			Industry		
		Coefficient on drought (t-value)	Adjusted R2 (F-stat)	Coefficient on drought (t-value)	Agriculture as % GDP	Coefficient on drought weighed by share of sector in GDP	Adjusted R2 (F-stat)	Coefficient on drought (t-value)	Coefficient on drought weighed by share of sector in GDP	Adjusted R2 (F-stat)	Coefficient on drought (t-value)	Coefficient on drought weighed by share of sector in GDP	Adjusted R2 (F-stat)
Burkina Faso	1971-80	0.51 (3.19)**	0.62 * (8.3)	0.69 (2.43)	53.112	0.36	0.67 ** (10.1)	0.02 (0.09)	0.00	0.00 (0.5)	0.27 (0.29)	0.06	0.00 (0.3)
	1980-90	0.24 (1.98)	0.22 (2.4)	0.73 (3.70) **	44.98	0.33	0.62 ** (9.1)	-0.29 (-0.96)	-0.04	0.12 (1.7)	-0.30 (-1.20)	-0.05	0.07 (1.4)
Ethiopia	1975/6-90/1	0.28 (2.74)**	0.50 ** (6.1)	0.47 (2.19) *	45.27	0.21	0.27 (2.8)	0.16 (0.83)	0.02	0.43 * (4.8)	0.11 (0.65)	0.02	0.48 * (5.5)
Kenya	1971-81	0.03 (0.29)	0.00 (0.1)	0.02 (0.24)	34.59	0.01	0.00 (0.1)	b	b	b	-0.07 (-0.30)	-0.01	0.00 (0.1)
	1981-89	0.04 (2.77)*	0.45 * (7.7)	0.07 (2.10) *	31.65	0.02	0.30 (4.4)	0.01 (0.72)	0.00	0.00 (0.5)	0.02 (0.95)	0.00	0.00 (0.9)
Senegal	1971-80	0.29 (3.93)**	0.62 ** (15.5)	0.92 (3.04) **	25.84	0.24	0.48 ** (9.2)	0.38 (1.90)	0.04	0.27 (3.6)	0.23 (1.41)	0.04	0.10 (2.0)
	1980-90	0.19 (2.23)*	0.28 (5.0)	0.46 (2.11) *	21.61	0.10	0.26 (4.5)	-0.01 (-0.12)	-0.00	0.00 (0.01)	0.27 (2.95) **	0.05	0.44 * (8.7)
Zambia	1971-81	0.03 (0.32)	0.00 (0.1)	0.15 (2.50) *	15.39	0.02	0.34 (6.25)	-0.07 (-0.54)	-0.01	0.00 (0.3)	0.02 (0.19)	0.01	0.00 (0.0)
	1981-92	0.09 (2.35)**	0.29 ** (5.5)	0.44 (3.39) **	16.82	0.07	0.49 ** (11.5)	0.07 (0.69)	0.02	0.00 (0.5)	0.00 (.04)	0.00	0.00 (0.0)
Zimbabwe	1971-79	-0.06 (-1.49)	0.75 ** (12.8)	0.08 (0.41)	18.21	0.01	0.54 (3.5)	-0.10 (-0.90)	-0.02	0.19 (1.9)	-0.16 (-2.2) *	-0.06	0.43 (4.0)
	1980-92	0.16 (5.34)**	0.71 ** (16.0)	0.36 (3.57) **	15.19	0.05	0.64 ** (11.4)	0.18 (3.99) **	0.05	0.54 ** (8.1)	0.12 (3.00) **	0.03	0.39 * (4.8)

a GDP at factor cost except in the cases of Senegal and Zambia. The results are based on regressions using the following drought variables, all in deviation below period mean level indices: Burkina Faso - FAO national rainfall index.; Ethiopia - maize, sorghum and teff yields weighted by their shares in output; Kenya - March-June rainfall at Nairobi (Kenyatta airport) rainfall station; Senegal - groundnut yields, lagged one year; Zambia - February rainfall af Mongu and Livingstone rainfall stations; and Zimbabwe - national February rainfall.

b Manufacturing GDP data not available for the full period of analysis

* F-test or t-test significant at 5% level of confidence (using one-tailed t-tests)

** F-test or t-test significant at 1% level of confidence (using one-tailed t-tests)

Source: Benson 1994

Appendix Figure 1: Burkina Faso, economic performance and rainfall

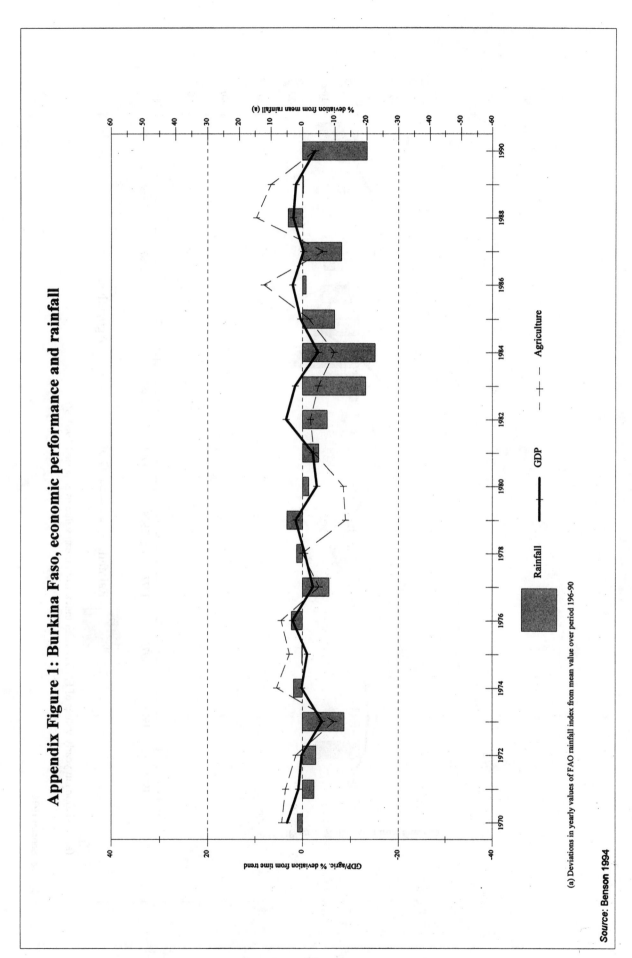

(a) Deviations in yearly values of FAO rainfall index from mean value over period 196-90

Source: Benson 1994

Appendix Figure 2: Ethiopia, economic performance and rainfall

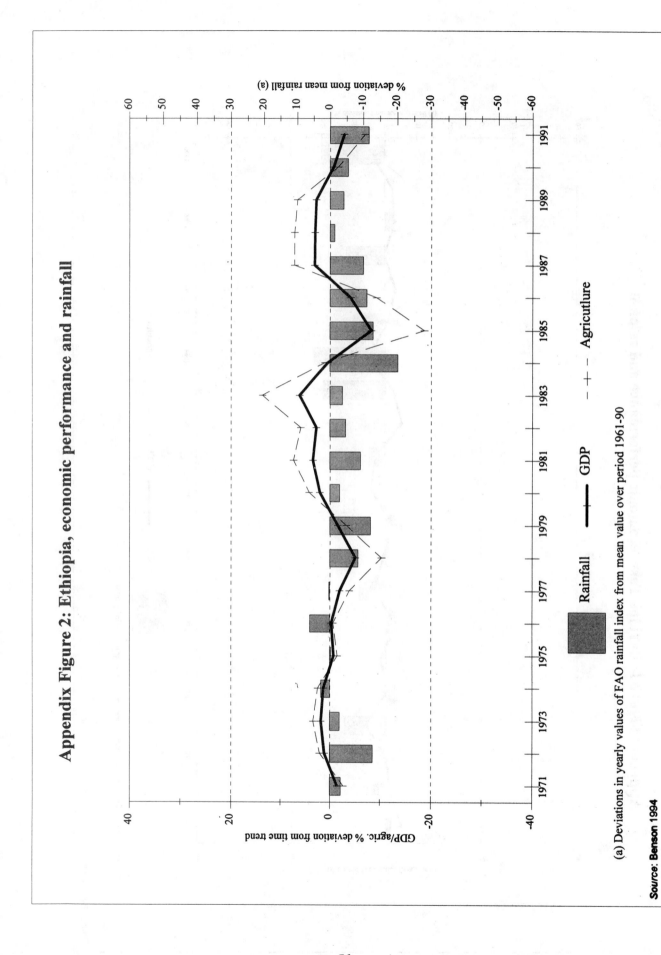

(a) Deviations in yearly values of FAO rainfall index from mean value over period 1961-90

Source: **Benson 1994**

76

Appendix Figure 3: Kenya, economic performance and rainfall

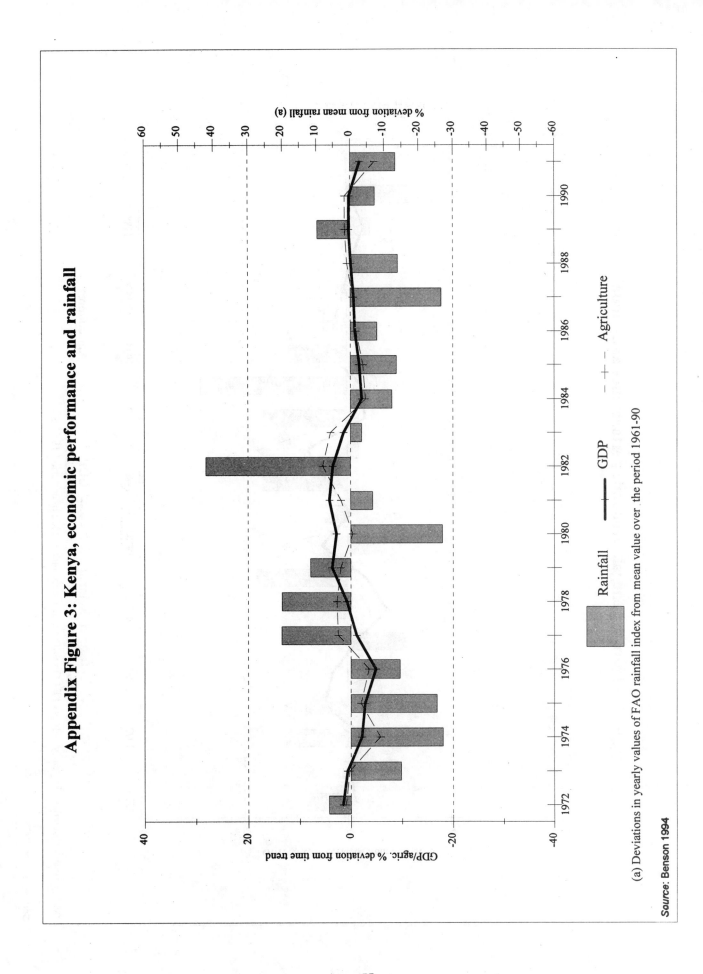

(a) Deviations in yearly values of FAO rainfall index from mean value over the period 1961-90

Source: Benson 1994

Appendix Figure 4: Senegal, economic performance and rainfall

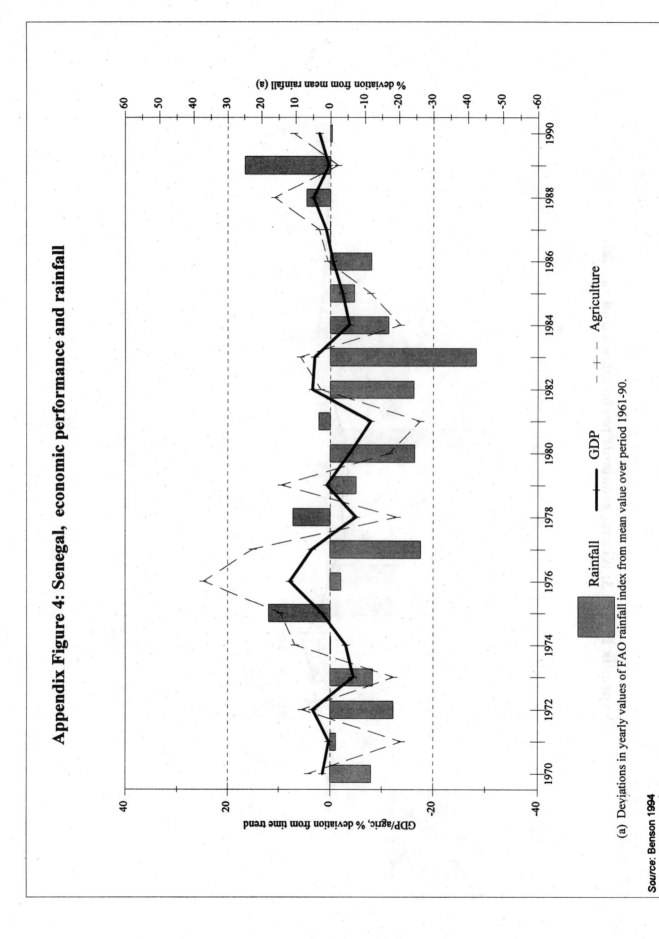

(a) Deviations in yearly values of FAO rainfall index from mean value over period 1961-90.

Source: Benson 1994

Appendix Figure 5: Zambia, economic performance and rainfall

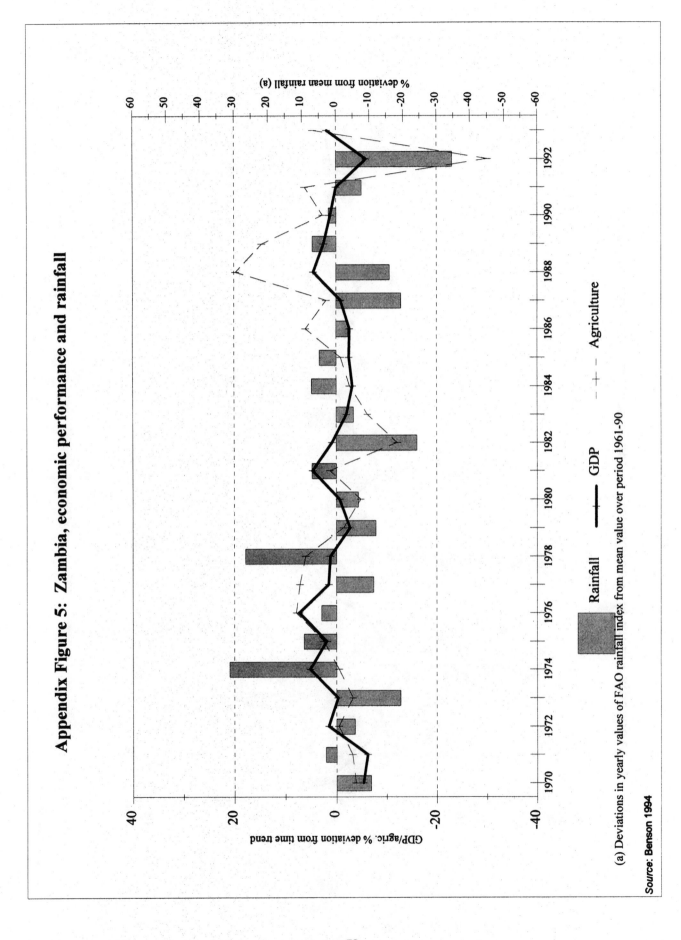

Rainfall   ── GDP   ─+─ Agriculture

(a) Deviations in yearly values of FAO rainfall index from mean value over period 1961-90

Source: **Benson 1994**

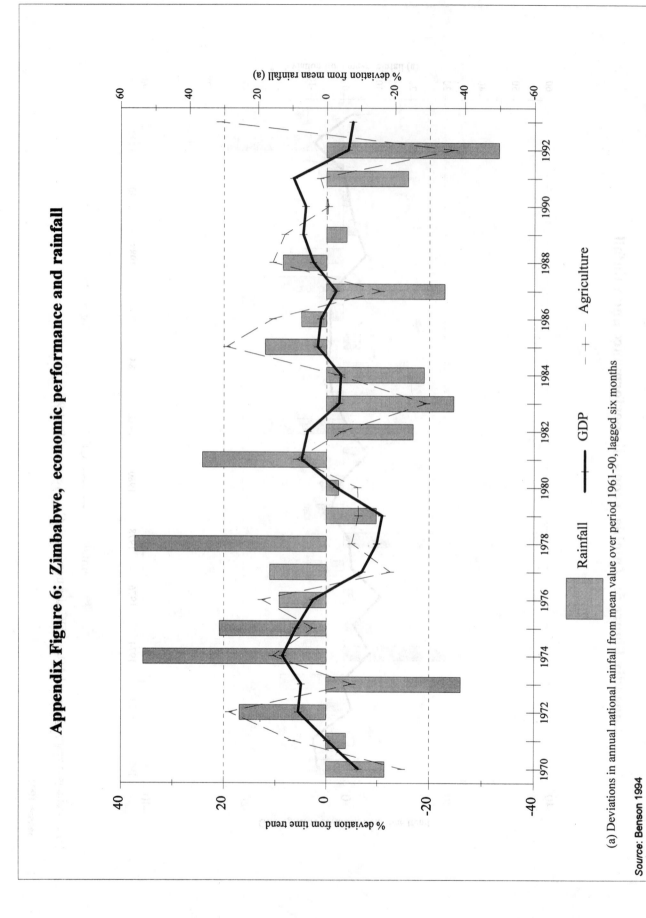

Appendix Figure 6: Zimbabwe, economic performance and rainfall

% deviation from mean rainfall (a)

% deviation from time trend

| | Rainfall | —— | GDP | -+- | Agriculture |

(a) Deviations in annual national rainfall from mean value over period 1961-90, lagged six months

Source: Benson 1994

80

Distributors of World Bank Publications

Prices and credit terms vary from country to country. Consult your local distributor before placing an order.

ARGENTINA
Oficina del Libro Internacional
Av. Cordoba 1877
1120 Buenos Aires
Tel: (54 1) 815-8354
Fax: (54 1) 815-8156

AUSTRALIA, FIJI, PAPUA NEW GUINEA, SOLOMON ISLANDS, VANUATU, AND WESTERN SAMOA
D.A. Information Services
648 Whitehorse Road
Mitcham 3132
Victoria
Tel: (61) 3 9210 7777
Fax: (61) 3 9210 7788
E-mail: service@dadirect.com.au
URL: http://www.dadirect.com.au

AUSTRIA
Gerold and Co.
Weihburggasse 26
A-1011 Wien
Tel: (43 1) 512-47-31-0
Fax: (43 1) 512-47-31-29
URL: http://www.gerold.co/at.online

BANGLADESH
Micro Industries Development
Assistance Society (MIDAS)
House 5, Road 16
Dhanmondi R/Area
Dhaka 1209
Tel: (880 2) 326427
Fax: (880 2) 811188

BELGIUM
Jean De Lannoy
Av. du Roi 202
1060 Brussels
Tel: (32 2) 538-5169
Fax: (32 2) 538-0841

BRAZIL
Publicações Tecnicas Internacionais Ltda.
Rua Peixoto Gomide, 209
01409 Sao Paulo, SP.
Tel: (55 11) 259-6644
Fax: (55 11) 258-6990
E-mail: postmaster@pti.uol.br
URL: http://www.uol.br

CANADA
Renouf Publishing Co. Ltd.
5369 Canotek Road
Ottawa, Ontario K1J 9J3
Tel: (613) 745-2665
Fax: (613) 745-7660
E-mail: order.dept@renoufbooks.com
URL: http://www.renoufbooks.com

CHINA
China Financial & Economic
Publishing House
8, Da Fo Si Dong Jie
Beijing
Tel: (86 10) 6333-8257
Fax: (86 10) 6401-7365

COLOMBIA
Infoenlace Ltda.
Carrera 6 No. 51-21
Apartado Aereo 34270
Santafé de Bogotá, D.C.
Tel: (57 1) 285-2798
Fax: (57 1) 285-2798

COTE D'IVOIRE
Center d'Edition et de Diffusion Africaines
(CEDA)
04 B.P. 541
Abidjan 04
Tel: (225) 24 6510;24 6511
Fax: (225) 25 0567

CYPRUS
Center for Applied Research
Cyprus College
6, Diogenes Street, Engomi
P.O. Box 2006
Nicosia
Tel: (357 2) 44-1730
Fax: (357 2) 46-2051

CZECH REPUBLIC
National Information Center
prodejna, Konviktska 5
CS – 113 57 Prague 1
Tel: (42 2) 2422-9433
Fax: (42 2) 2422-1484
URL: http://www.nis.cz/

DENMARK
SamfundsLitteratur
Rosenoerns Allé 11
DK-1970 Frederiksberg C
Tel: (45 31) 357822
Fax: (45 31) 357822
URL: http://www.sl.cbs.dk

ECUADOR
Libri Mundi
Libreria Internacional
P.O. Box 17-01-3029
Juan Leon Mera 851
Quito
Tel: (593 2) 521-606; (593 2) 544-185
Fax: (593 2) 504-209
E-mail: librimu1@librimundi.com.ec
E-mail: librimu2@librimundi.com.ec

EGYPT, ARAB REPUBLIC OF
Al Ahram Distribution Agency
Al Galaa Street
Cairo
Tel: (20 2) 578-6083
Fax: (20 2) 578-6833

The Middle East Observer
41, Sherif Street
Cairo
Tel: (20 2) 393-9732
Fax: (20 2) 393-9732

FINLAND
Akateeminen Kirjakauppa
P.O. Box 128
FIN-00101 Helsinki
Tel: (358 0) 121 4418
Fax: (358 0) 121-4435
E-mail: akatilaus@stockmann.fi
URL: http://www.akateeminen.com/

FRANCE
World Bank Publications
66, avenue d'Iéna
75116 Paris
Tel: (33 1) 40-69-30-56/57
Fax: (33 1) 40-69-30-68

GERMANY
UNO-Verlag
Poppelsdorfer Allee 55
53115 Bonn
Tel: (49 228) 949020
Fax: (49 228) 217492
URL: http://www.uno-verlag.de
E-mail: unoverlag@aol.com

GREECE
Papasotiriou S.A.
35, Stournara Str.
106 82 Athens
Tel: (30 1) 364-1826
Fax: (30 1) 364-8254

HAITI
Culture Diffusion
5, Rue Capois
C.P. 257
Port-au-Prince
Tel: (509) 23 9260
Fax: (509) 23 4858

HONG KONG, MACAO
Asia 2000 Ltd.
Sales & Circulation Department
Seabird House, unit 1101-02
22-28 Wyndham Street, Central
Hong Kong
Tel: (852) 2530-1409
Fax: (852) 2526-1107
E-mail: sales@asia2000.com.hk
URL: http://www.asia2000.com.hk

HUNGARY
Euro Info Service
Margitszgeti Europa Haz
H-1138 Budapest
Tel: (36 1) 111 6061
Fax: (36 1) 302 5035
E-mail: euroinfo@mail.matav.hu

INDIA
Allied Publishers Ltd.
751 Mount Road
Madras - 600 002
Tel: (91 44) 852-3938
Fax: (91 44) 852-0649

INDONESIA
Pt. Indira Limited
Jalan Borobudur 20
P.O. Box 181
Jakarta 10320
Tel: (62 21) 390-4290
Fax: (62 21) 390-4289

IRAN
Ketab Sara Co. Publishers
Khaled Eslamboli Ave., 6th Street
Delafrooz Alley No. 8
P.O. Box 15745-733
Tehran 15117
Tel: (98 21) 8717819; 8716104
Fax: (98 21) 8712479

Kowkab Publishers
P.O. Box 19575-511
Tehran
Tel: (98 21) 258-3723
Fax: (98 21) 258-3723

IRELAND
Government Supplies Agency
Oifig an tSoláthair
4-5 Harcourt Road
Dublin 2
Tel: (353 1) 661-3111
Fax: (353 1) 475-2670

ISRAEL
Yozmot Literature Ltd.
P.O. Box 56055
3 Yohanan Hasandlar Street
Tel Aviv 61560
Tel: (972 3) 5285-397
Fax: (972 3) 5285-397

R.O.Y. International
PO Box 13056
Tel Aviv 61130
Tel: (972 3) 5461423
Fax: (972 3) 5461442
E-mail: royil@netvision.net.il

Palestinian Authority/Middle East
Index Information Services
P.O.B. 19502 Jerusalem
Tel: (972 2) 6271219
Fax: (972 2) 6271634

ITALY
Licosa Commissionaria Sansoni SPA
Via Duca Di Calabria, 1/1
Casella Postale 552
50125 Firenze
Tel: (55) 645-415
Fax: (55) 641-257
E-mail: licosa@ftbcc.it
URL: http://www.ftbcc.it/licosa

JAMAICA
Ian Randle Publishers Ltd.
206 Old Hope Road, Kingston 6
Tel: 876-927-2085
Fax: 876-977-0243
E-mail: irpl@colis.com

JAPAN
Eastern Book Service
3-13 Hongo 3-chome, Bunkyo-ku
Tokyo 113
Tel: (81 3) 3818-0861
Fax: (81 3) 3818-0864
E-mail: orders@svt-ebs.co.jp
URL: http://www.bekkoame.or.jp/~svt-ebs

KENYA
Africa Book Service (E.A.) Ltd.
Quaran House, Mfangano Street
P.O. Box 45245
Nairobi
Tel: (254 2) 223 641
Fax: (254 2) 330 272

KOREA, REPUBLIC OF
Daejon Trading Co. Ltd.
P.O. Box 34, Youida, 706 Seoun Bldg
44-6 Youido-Dong, Yeongchengpo-Ku
Seoul
Tel: (82 2) 785-1631/4
Fax: (82 2) 784-0315

MALAYSIA
University of Malaya Cooperative
Bookshop, Limited
P.O. Box 1127
Jalan Pantai Baru
59700 Kuala Lumpur
Tel: (60 3) 756-5000
Fax: (60 3) 755-4424
E-mail: umkoop@tm.net.my

MEXICO
INFOTEC
Av. San Fernando No. 37
Col. Toriello Guerra
14050 Mexico, D.F.
Tel: (52 5) 624-2800
Fax: (52 5) 624-2822
E-mail: infotec@rtn.net.mx
URL: http://rtn.net.mx

NEPAL
Everest Media International Services (P) Ltd.
GPO Box 5443
Kathmandu
Tel: (977 1) 472 152
Fax: (977 1) 224 431

NETHERLANDS
De Lindeboom/InOr-Publikaties
P.O. Box 202, 7480 AE Haaksbergen
Tel: (31 53) 574-0004
Fax: (31 53) 572-9296
E-mail: lindeboo@worldonline.nl
URL: http://www.worldonline.nl/~lindeboo

NEW ZEALAND
EBSCO NZ Ltd.
Private Mail Bag 99914
New Market
Auckland
Tel: (64 9) 524-8119
Fax: (64 9) 524-8067

NIGERIA
University Press Limited
Three Crowns Building Jericho
Private Mail Bag 5095
Ibadan
Tel: (234 22) 41-1356
Fax: (234 22) 41-2056

NORWAY
NIC Info A/S
Book Department, Postboks 6512 Etterstad
N-0606 Oslo
Tel: (47 22) 97-4500
Fax: (47 22) 97-4545

PAKISTAN
Mirza Book Agency
65, Shahrah-e-Quaid-e-Azam
Lahore 54000
Tel: (92 42) 735 3601
Fax: (92 42) 576 3714

Oxford University Press
5 Bangalore Town
Sharae Faisal
PO Box 13033
Karachi-75350
Tel: (92 21) 446307
Fax: (92 21) 4547640
E-mail: ouppak@TheOffice.net

Pak Book Corporation
Aziz Chambers 21, Queen's Road
Lahore
Tel: (92 42) 636 3222; 636 0885
Fax: (92 42) 636 2328
E-mail: pbc@brain.net.pk

PERU
Editorial Desarrollo SA
Apartado 3824, Lima 1
Tel: (51 14) 285380
Fax: (51 14) 286628

PHILIPPINES
International Booksource Center Inc.
1127-A Antipolo St, Barangay, Venezuela
Makati City
Tel: (63 2) 896 6501; 6505; 6507
Fax: (63 2) 896 1741

POLAND
International Publishing Service
Ul. Piekna 31/37
00-677 Warzawa
Tel: (48 2) 628-6089
Fax: (48 2) 621-7255
E-mail: books%ips@ikp.atm.com.pl
URL: http://www.ipscg.waw.pl/ips/export/

PORTUGAL
Livraria Portugal
Apartado 2681, Rua Do Carmo 70-74
1200 Lisbon
Tel: (1) 347-4982
Fax: (1) 347-0264

ROMANIA
Compani De Librarii Bucuresti S.A.
Str. Lipscani no. 26, sector 3
Bucharest
Tel: (40 1) 613 9645
Fax: (40 1) 312 4000

RUSSIAN FEDERATION
Isdatelstvo <Ves Mir>
9a, Kolpachniy Pereulok
Moscow 101831
Tel: (7 095) 917 87 49
Fax: (7 095) 917 92 59

SINGAPORE, TAIWAN, MYANMAR, BRUNEI
Ashgate Publishing Asia Pacific Pte. Ltd.
41 Kallang Pudding Road #04-03
Golden Wheel Building
Singapore 349316
Tel: (65) 741-5166
Fax: (65) 742-9356
E-mail: ashgate@asianconnect.com

SLOVENIA
Gospodarski Vestnik Publishing Group
Dunajska cesta 5
1000 Ljubljana
Tel: (386 61) 133 83 47; 132 12 30
Fax: (386 61) 133 80 30
E-mail: repansekj@gvestnik.si

SOUTH AFRICA, BOTSWANA
For single titles:
Oxford University Press Southern Africa
Vasco Boulevard, Goodwood
P.O. Box 12119, N1 City 7463
Cape Town
Tel: (27 21) 595 4400
Fax: (27 21) 595 4430
E-mail: oxford@oup.co.za

For subscription orders:
International Subscription Service
P.O. Box 41095
Craighall
Johannesburg 2024
Tel: (27 11) 880-1448
Fax: (27 11) 880-6248
E-mail: iss@is.co.za

SPAIN
Mundi-Prensa Libros, S.A.
Castello 37
28001 Madrid
Tel: (34 1) 431-3399
Fax: (34 1) 575-3998
E-mail: libreria@mundiprensa.es
URL: http://www.mundiprensa.es/

Mundi-Prensa Barcelona
Consell de Cent, 391
08009 Barcelona
Tel: (34 3) 488-3492
Fax: (34 3) 487-7659
E-mail: barcelona@mundiprensa.es

SRI LANKA, THE MALDIVES
Lake House Bookshop
100, Sir Chittampalam Gardiner Mawatha
Colombo 2
Tel: (94 1) 32105
Fax: (94 1) 432104
E-mail: LHL@sri.lanka.net

SWEDEN
Wennergren-Williams AB
P.O. Box 1305
S-171 25 Solna
Tel: (46 8) 705-97-50
Fax: (46 8) 27-00-71
E-mail: mail@wwi.se

SWITZERLAND
Librairie Payot Service Institutionnel
Côtes-de-Montbenon 30
1002 Lausanne
Tel: (41 21) 341-3229
Fax: (41 21) 341-3235

ADECO Van Diermen EditionsTechniques
Ch. de Lacuez 41
CH1807 Blonay
Tel: (41 21) 943 2673
Fax: (41 21) 943 3605

THAILAND
Central Books Distribution
306 Silom Road
Bangkok 10500
Tel: (66 2) 235-5400
Fax: (66 2) 237-8321

TRINIDAD & TOBAGO AND THE CARRIBBEAN
Systematics Studies Ltd.
St. Augustine Shopping Center
Eastern Main Road, St. Augustine
Trinidad & Tobago, West Indies
Tel: (868) 645-8466
Fax: (868) 645-8467
E-mail: tobe@trinidad.net

UGANDA
Gustro Ltd.
PO Box 9997, Madhvani Building
Plot 16/4 Jinja Rd.
Kampala
Tel: (256 41) 251 467
Fax: (256 41) 251 468
E-mail: gus@swiftuganda.com

UNITED KINGDOM
Microinfo Ltd.
P.O. Box 3, Alton, Hampshire GU34 2PG
England
Tel: (44 1420) 86848
Fax: (44 1420) 89889
E-mail: wbank@ukminfo.demon.co.uk
URL: http://www.microinfo.co.uk

VENEZUELA
Tecni-Ciencia Libros, S.A.
Centro Cuidad Comercial Tamanco
Nivel C2, Caracas
Tel: (58 2) 959 5547; 5035; 0016
Fax: (58 2) 959 5636

ZAMBIA
University Bookshop, University of Zambia
Great East Road Campus
P.O. Box 32379
Lusaka
Tel: (260 1) 252 576
Fax: (260 1) 253 952

10/0997